Design Science and Innovation

Series Editor

Amaresh Chakrabarti, Centre for Product Design and Manufacturing, Indian Institute of Science, Bangalore, India

The book series is intended to provide a platform for disseminating knowledge in all areas of design science and innovation, and is intended for all stakeholders in design and innovation, e.g. educators, researchers, practitioners, policy makers and students of design and innovation. With leading international experts as members of its editorial board, the series aims to disseminate knowledge that combines academic rigour and practical relevance in this area of crucial importance to the society.

More information about this series at https://link.springer.com/bookseries/15399

Jegatheswaran Ratnasingam

Furniture Manufacturing

A Production Engineering Approach

Springer

Jegatheswaran Ratnasingam
Faculty of Forestry and Environment
Universiti Putra Malaysia
Seri Kembangan, Selangor, Malaysia

ISSN 2509-5986 ISSN 2509-5994 (electronic)
Design Science and Innovation
ISBN 978-981-16-9411-0 ISBN 978-981-16-9412-7 (eBook)
https://doi.org/10.1007/978-981-16-9412-7

This Springer imprint is published by the registered company Springer Nature Singapore Pte Ltd.
The registered company address is: 152 Beach Road, #21-01/04 Gateway East, Singapore 189721, Singapore

Dedicated to all great furniture makers of the past and present

Preface

The impetus for this book came from the realization that professional books on furniture manufacturing processes are very limited, and the few available takes an artisan or craftsman approach. Although several publications on woodworking which takes the vocational-education perspective are available, it is not suited for university level and professional practitioners, as it lacks the depth and consolidated approach required of such a publication. In fact, there is overwhelming information on woodworking for hobbyist and individual woodworker on the World Wide Web, but such information cannot find application for university-level education. Further my years of teaching and research at the Technical Hochschule Rosenheim in Germany, regarded the leading woodworking institute of higher education in the world, from 1999 until 2019, impressed upon me that any curriculum on furniture manufacturing and wood technology at large, must assume an engineering approach in order to remain viable in this globalized world, where technology impinges on all aspects of the manufacturing industry. In fact, this has been the singular reason for the sustainability and viability of wood science and technology education at large, in Germany and a few central European countries, when throughout the world, many such programs in the United States of America, the United Kingdom, Australia, Japan, etc., have closed due to dwindling student enrollment.

It took me several years to refine the thoughts and convert it into a book project plan. As I began to solidify the concept, *Professor Emeritus György Sitkei* of the University of West Hungary invited me to participate as an editor of the book "*Optimum Design and Manufacture of Wood Products*" published by Springer-Nature in 2019. The chapter on furniture manufacturing processes published in that book served as the precursor for the thoroughly expanded and comprehensive book on furniture manufacturing. At the same time, much of the material presented in this book relied on my own research work. At this juncture, I wish to state that the figures, tables, and illustrations contained in this book are from my own collection, unless otherwise mentioned.

The overall goal of this book is twofold. First, I want to provide a means to support and advance teaching in furniture manufacturing fields, along the production engineering perspective. Second, I want to consolidate current knowledge on various themes in furniture manufacturing that would be useful to established and emerging scholars, industry personnel, and policy makers. Extant knowledge in furniture manufacturing has largely remained dispersed across disparate academic sources albeit comprehensive books are available for specific areas such as wood science and technology, wood adhesives, wood coatings, and processing technologies. Thus, a gap was identified in the literature that was preventing students, scholars, policymakers, and others from developing a timely, structured, big-picture view of furniture manufacturing. Furthermore, there was no single resource available for reviewing the current state of the art and trends on the many aspects of furniture manufacturing. This gap looked even wider and deeper given the dramatic changes that have taken place in the furniture sector as a whole since the global financial crisis of the last decade, the advent of the circular and digital economies, as well as the rapid expansion in manufacturing technologies based on Industry 4.0. The fact that the existing publication on the subject of furniture manufacturing processes by *Prak and Myers* (1978) assumes a management perspective rather an engineering one, provided further encouragement for the writing of this comprehensive book with an elaboration of all aspects of furniture manufacturing.

How successful will this book be in filling this gap is for the reader to judge, but I feel this book is an important step towards the integration of further development of the furniture manufacturing field as a sub-sector of production engineering. This book could be used to support an entire academic course on furniture manufacturing. Individual parts may be useful as supplemental readings for other courses at both graduate and undergraduate levels. Also, because this book captures concepts from a broad swath of the furniture manufacturing industry, it can help factory managers and policy makers to update their knowledge base, especially in areas which they are less familiar or outside their core responsibilities.

In the world of furniture manufacturing, I would have liked to write a comprehensive book and title it "Furniture Manufacturing Handbook," but the innumerous topics, many of which are highly technical and subject-specific, precluded their inclusion in a book of this nature. Whatever the reasons for those omissions, I sincerely hope that colleagues in academia, research organizations, industry practitioners, and emerging scholars will find this book useful as a pedagogical tool, as a basis to identify research areas and fill research and information gaps to make the discipline of furniture manufacturing technology and management even more interesting, richer, and relevant to changing times.

Finally, I am also sincerely grateful to the staff at Springer-Nature Publishing, especially Ms. Muskan Jaiswal and Ms. Sushmitha S. Sundaram for their support, assistance, and excellent coordination during the course of publishing this book.

Happy reading.

Serdang, Malaysia
October 2021

Jegatheswaran Ratnasingam

Acknowledgements

Books such as this are successful to the extent, they receive inputs from many scholars and industry experts who are highly respected for their knowledge within their field of disciplines. I cannot thank enough each individual who contributed to the creation of this book. The support I received from colleagues and industry experts in fact, echoes the need for this book. However, special acknowledgement must be made to the contributions of Emeritus Professor György Sitkei of the University of West Hungary, Germany, Emeritus Professor Dr. Carl Eckelman of Purdue University, United States of America, and Professor Dr. Frieder Scholz of the Technical Hochschule Rosenheim in Germany for their mentorship, extensive discussion, constructive comments, and useful insights that brought this book to its final form.

My scientific pursuit of furniture manufacturing and management began while studying for my Doctoral degree at the Buckinghamshire College in High Wycombe, United Kingdom, under the tutelage of Professor Dr. Hew F. Reid and Dr. Marius C. Perkins. I am grateful to them for starting me off as a researcher cum academic in the furniture field. Acknowledgement is also due to the Malaysian Furniture Council (MFC) and the Confederation of Asian Furniture Associations (CAFA) for their support in sharing relevant trade and market information, and also many figures used in this book. I am also indebted to my colleagues at the Food and Agriculture Organization (FAO) in Rome, International Trade Center (ITC) in Geneva, and the International Tropical Timber Organization (ITTO) in Yokohama for their comments and inputs.

Writing a book of this nature, is enduring and taxing, and without the support of the family it would not have been possible. In this respect, I thank my wife, Dr. Vidya and my two sons, Natkuncaran and Ayenkaren for tolerating my incessant preoccupation during the writing of this book. A happy and supportive family makes both the journey and destination worthwhile.

Finally, as with all my writings, my father, the late Mr. Ratnasingam and my mother, Mrs. Rajeswari, remain the focus in my mind's eye. Their tenacity for knowledge and wonder has been pivotal in shaping my enquiring mind. Everything I have and everything I am, I owe it all to both of you. Thank you, my two lifelines.

Contents

About the Author

Jegatheswaran Ratnasingam, Faculty of Forestry & Environment, Universiti Putra Malaysia, 43400 UPM, Serdang, Selangor, Malaysia.

He was involved in the furniture manufacturing industry throughout Asia in various capacities, for almost one and a half decade, before venturing into academia. He is currently the Professor of Furniture Manufacturing and Management at the Faculty of Forestry & Environment, at Universiti Putra Malaysia. As a leading researcher and academic in the field of furniture manufacturing technology and management, he has more than 550 publications to date, in scientific journals, conference proceedings, trade journals, reports, and books. He is a much sought-after industry consultant and symposia speaker both nationally and internationally, and has been on assignments with many international agencies throughout the world. In recognition of his contributions to the furniture industry, he has been awarded three Honorary Professorships from distinguished universities in the United Kingdom, Germany, and South Africa, and is a recipient of many other scientific and industry awards. He also holds advisory positions in several national furniture industry trade bodies, both in Malaysia and internationally.

List of Figures

List of Tables

Chapter 1
The Furniture Manufacturing Industry

Furniture being fashion and merchandise, is created based on designs, which are converted into tangible products, using the necessary inputs and technologies. Incepted as a cottage-based industry, it has grown to become multi-billion-dollar industries in many countries throughout the world, serving both the export and domestic market. It is predominated by small- and medium-sized enterprises, although the supply chain and value chain in the industry is increasingly intertwined into a complex network, rendering it flexibility in design and volume production. This chapter will provide insights into the development of the furniture industry, and its important characteristics.

1.1 Furniture as a Product

Furniture is movable objects used in various human activities. Chairs, stools, sofas, tables, beds, outdoor furniture, etc., are among the common furniture encountered in daily human life. Apart from its functional role of holding objects at a convenient work height, or to store things, furniture is also considered fashion and a work of art. In some instances, furniture also serves as religious symbols. In essence, furniture is often designed to play both functional and esthetical roles. Furniture can be produced from many different types of materials, including wood, wood-based composites, metal, plastic, rattan, bamboo, glass, etc.

Unlike other types of wood products, furniture is merchandise and fashion, and its value is defined by its perceived value, and not exactly its exact value (Ratnasingam 2015). The distinct difference between commodity-type wood products, such as sawn timber, plywood, particleboard, etc., and merchandise, such as furniture must be highlighted, as the value of furniture is not dictated by supply–demand conditions, but rather the market perception. It must also be emphasized that furniture as fashion is an identity of the lifestyle, which imparts pride of ownership and social status to the owners (MFC 2017).

J. Ratnasingam, *Furniture Manufacturing*, Design Science and Innovation,
https://doi.org/10.1007/978-981-16-9412-7_1

1.2 Furniture Manufacturing as a Global Industry

In 2019, the global furniture production was valued at USD580 billion, and it is expected to grow by 3.2% annually from 2020 to 2026 (CSIL 2019). The demand for furniture in the world market is fueled by the construction sector, or rather housing start-ups, and for replacement purchases. Further, the purchasing trend of furniture is impacted by the disposable income available, and therefore, countries with higher gross national product (GDP) tend to consume more furniture and vice-versa. Fully furnished commercial and residential buildings are more attractive to generation-Z and millennials, which in turn creates more demand for furniture as the construction of such buildings increases. The country's development has strong bearing on all aspects of its economy, and thus prosperity. A growing middle-class population, together with higher incomes, are good indicators of economic prosperity of the country (IFRG 2019). Good examples of such countries are China and India.

The global furniture industry is constantly looking at ways of producing furniture that suits different lifestyles of its customers. Technological advancements are also helping furniture manufacturers produce innovative and functional products which conveniently serves the day to day needs of their customers (CSIL 2019).

The amount of disposable income, changing lifestyles, and rapid urbanization in many countries contributes to the increasing demand for furniture globally. Trade practices that encompass reductions in tariffs on furniture trade, growing presence of global furniture retail chains, expanding supply chains involving foreign suppliers, and large-scale distributors (e.g., involving contract manufacturers and retailers) and improved logistics and transportation continue to take the global furniture business to new heights (Schuler and Buehlmann 2003; IFRG 2019). Indeed, a revolution of the furniture industry has taken place, whereby retailers are working on designs and marketing, while manufacturers are focusing on production activities.

According to a report by IFRG (2019), nearly 20% of the global furniture production is provided for by 200 major furniture manufacturing companies, with a worldwide presence. Almost 75% of the companies are based in the developed countries, while the balance 25% are in developing countries. The globalized furniture industry systemizes its design, production, distribution, and marketing through a complex network of supply chain. Among the notable furniture manufacturing companies in the world include, Herman Miller Inc., The Home Depot, Inc, Renaissance Furniture, Global Furniture Group, Heritage Home, Inter Kohler Co., La-Z-Boy, IKEA AB, Okamura Corporation, Ashley Furniture Industries, Inc. Furniture Concepts, Haworth, Inc, HNI Corporation, Godrej & Boyce Manufacturing Co., Urban Office Interiors, Humanscale Corporation, McCarthy Group, Furniture Services Inc., Steelcase Inc., Tempur-Sealy International Inc., Leggett & Platt Incorporated (IFRG 2019; CSIL 2019).

The report by IFRG (2019) also revealed that wood resources was the primary raw material for furniture production worldwide and held a 60% market share in 2019. Further, with the growing environmental consciousness and demand for sustainable forestry practices, wood resources are projected to remain the predominant furniture-making raw material throughout the world. Nevertheless, other materials including composites, plastic, metal, bamboo, rattan, and glass are becoming increasingly important. These materials are increasingly popular as the demand for lightweight and durable furniture grows worldwide, apart from specific advantages related to strength, cost, and flexibility that they offer.

Against this background, the main growth drivers of the furniture industry worldwide are: (1) changes in consumer behavior due to transformation in lifestyles, (2) increasing housing start-ups, and (3) increasing replacement purchases due to higher proportion of disposable income. On the other hand, the challenges faced by the furniture industry are: (1) unsustainable supply of raw materials, (2) limited availability of skilled labor, (3) insufficient investments into automation and technology, and (4) the need to cope with a complex network of supply chain (Bullard and West 2002; Ratnasingam and Ioras 2003; Drayse 2011).

1.3 Global Furniture Market

According to the report by CSIL (2019), the global furniture market is forecasted to decline from USD486 billion in 2019 to USD435 billion in 2020, due to the uncertainties in the global economy due to the COVID-19 pandemic. The impact of the COVID-19 pandemic on the global economy is severe and a global economic slowdown of proportions not experienced previously, is inevitable. Many countries throughout the world have introduced stimulus packages to restart their respective economies, and the market for furniture is projected to grow slowly to reach USD610 billion by 2025 (IFRG 2019).

The global trade of furniture has been steadily growing over the past decade and contributed 1.5% to the total trade of manufactured goods worldwide (CSIL 2019). Asia Pacific has emerged as the largest furniture manufacturing region in the world accounting for almost 49% of the global production in 2019. North America remains the second largest furniture producing region at 25% global market share, and followed by Europe as the third largest furniture producing region. In terms of furniture consumption, Africa remains the smallest region in the world. Residential furniture or home furnishing segments commands the highest market share of 63% among all types of furniture produced, and this trend is anticipated to continue for many more years to come. Increasing demand for luxury items, including high-end furniture driven by higher disposable income and changing lifestyles among the middle-class population in China and India is making the Asia Pacific region and important furniture market globally (Ratnasingam and Ioras 2005). With the rapid pace of globalization, some furniture producing nations have embraced technology and automation to transform their industries to remain competitive. Among the leading furniture producing countries in the world are as shown in Fig. 1.1.

Presently, China is the largest furniture producer in the world, which has a runaway lead compared to all other furniture producing nations. Other leading furniture

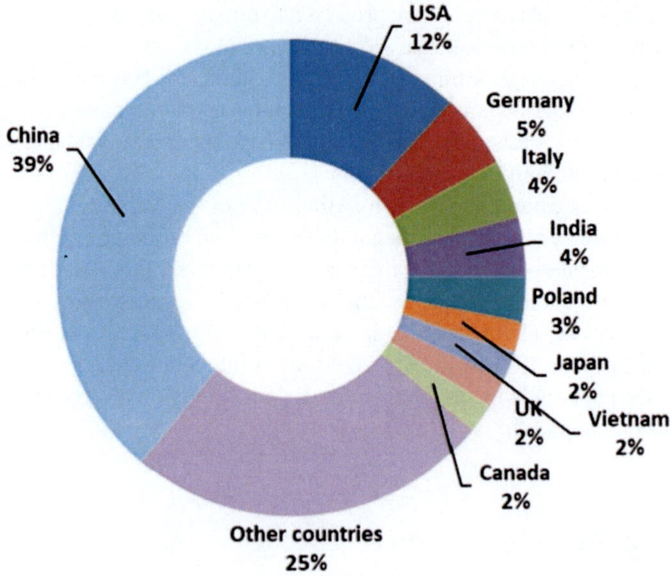

Fig. 1.1 Major Furniture Producing Countries, 2019 (*Source* CSIL)

producing nations in the order of reducing importance are the USA, Germany, Italy, India, Poland, Japan, Vietnam, the United Kingdom, and Canada. Since the 1990s, China's unprecedented growth in its furniture production capacity has made it the furniture factory of the world, and in 2019, its total production was valued at USD86 billion (CSIL 2019). Factors such as the export-driven industrial policy, increasing investments (of both domestic and foreign), and low-cost factor inputs (including raw material and labor) have contributed to the rapid rise of China as the leading furniture producer in the world. It must however be recognized that almost 65% of its production capacity is meant for its domestic consumption, fueled by rapid urbanization, and increasing middle-class population (Ratnasingam 2015).

For decades, the USA has been regarded as the furniture capital of world, due to her undisputable position as the leading center of furniture production and trade. However, with more and more furniture retailers importing the supplies from cheaper manufacturers in the Asia–Pacific region, especially from China and Vietnam, domestic production of furniture in the USA declined steadily since the early 2000s (Schuler and Buehlmann 2003; Quesada and Gazo 2006). An unprecedented number of furniture factories closed due to poor domestic demand and the stiff competition from the imports. Consequently, the decline of furniture manufacturing in the USA, accelerated the rise of China as the furniture factory of the world.

The rise of generation-X and millennial consumers have also brought about a transformed furniture business. Brand consciousness is gaining traction among these consumer groups, and inevitably, furniture manufacturers must move along further the value-chain to be able to manufacture branded products. The rapid expansion of e-commerce and digital marketing platforms, sales of furniture globally have also been encouraged further. To cater for the millennials, furniture retailers are increasingly focused on offering product diversity based on the do-it-yourself (DIY) concept, which ensures product affordability, and market growth (CSIL 2019). Although furniture serves functional roles, increasingly consumers are buying furniture for home décor purposes, which in turn creates new market growth across the globe. The rapidly expanding construction of residential and commercial buildings in the Asia–Pacific is the salient factor that paves the way for an increasing number of furniture manufacturers who are exploring the opportunities in the shop-fit and house-fit furniture manufacturing. (IFRG 2019). In this regard, many furniture manufacturers in Asia–Pacific have launched their own online digital platforms, apart from making partnerships with e-commerce retail stores to tap into this market segment.

Since the onset of the global COVID-19 pandemic in early 2019, the demand for multifunctional furniture has risen significantly. Furniture that is stylish but space-saving with sufficient storage is highly desirable. To cope with the new normal and working from home, demand for small office, home office (SOHO) furniture has been exponential growth (Ratnasingam et al. 2020). SOHO furniture must be produced to the preferences and taste of customers, and hence, product flexibility has become an integral requirement of these furniture. Nevertheless, ready-to-assemble (RTA) or flat-packed furniture remains popular globally among single-home owners, renters, and students due to its lower costs and compact designs (Ratnasingam et al. 2017).

Despite the overall shift in furniture production towards low-cost countries globally, there is evidence to suggest that furniture manufacturing remains to be an important socioeconomic sector in many of the traditional furniture producing nations worldwide (Ratnasingam 2020). Furniture manufacturing in Italy continues to strive on the background of its fashion-oriented production, while Germany offsets its high labor cost through the adoption of technology and automation. Poland, a relatively newcomer in the global furniture production scene has enjoyed rapid expansion fueled by large foreign investments and a system-oriented approach to furniture manufacturing. Taiwan with its long tradition in furniture manufacturing continue to strive through complex industrial clusters, which has enabled it to duplicate such models in China and Vietnam effectively, inevitably contributing to the rapid growth of the furniture industry in these countries (Ratnasingam 2020).

1.4 Development of the Furniture Industry: A Case Study of Malaysia

The furniture manufacturing industry is considered a traditional and labor-intensive industry. From its humble beginning as a cottage industry in the 1980s, offering products to the domestic market, the furniture industry has grown to become multi-billion-dollar export-oriented industry over a period of three decades. Malaysia is presently among the top 15 largest exporters of furniture in the world, with exports to more than 116 countries throughout the world. The main markets for Malaysian furniture include the USA, Japan, the United Kingdom, Europe, Singapore, India, Australia, and the Middle East. In a recent survey published, it was shown that Malaysian furniture is highly sought after by customers throughout the world for its good workmanship, quality, and affordability (IFRG 2019).

The growth of the Malaysian furniture industry was kickstarted with the recognition of rubberwood (*Hevea brasiliensis*) as an alternative wood resource for furniture manufacturing. The intensive research program by the Forest Research Institute of Malaysia (FRIM) and the Malaysian Timber Industry Board (MTIB), had managed to convert the waste wood into a valuable wood resource (Ratnasingam 2015). With this newly found wood resource, and the availability of a large pool of workforce, the furniture industry grew rapidly. Industrial growth was accelerated with the participation of foreign investors from Taiwan, South Korea, and Singapore, who relocated their manufacturing plants to the shores of Malaysia to take advantage of these low-cost factor inputs (Ratnasingm 2015). From an international perspective, the breakdown in the furniture supply chain, from Yugoslavia to the USA due to the war, created a market opportunity for Malaysian exporters. The rapid growth of the industry was also supported through the implementation of a series of Industrial Master Plans (IMPs), i.e., first IMP (1986–1995), second IMP (1996–2005), third IMP (2006–2020), and the National Timber Industry Plan (NATIP) which transformed the industry into the leading manufacturer of furniture and other value-added wood products. The creation of a business-friendly environment through policy, regulatory and fiscal instruments made Malaysia an attractive destination for foreign investors in the furniture sector (NATIP 2009). Against this background, it is no surprise that the Malaysian furniture industry has become a model to be emulated by many aspiring furniture manufacturing nations throughout the world.

In 2019, Malaysia exported USD2.58 billion worth of furniture, with wooden furniture contributing 83% of the total. The domestic market for furniture in the country was valued at USD2.3 billion, with wooden furniture contributing USD0.4 billion, while the balance were made up of other types of furniture. This positive trend in furniture consumption is expected to grow further in years to come. The industry also creates spill-over benefits in the range of USD1.4 billion through other indirect and supporting activities. Further, the furniture industry provides direct employment to 105,000 people, and if the indirect employment is considered the total would grow to 220,000 people. Against this background, it is no surprise that the Malaysian furniture industry has become an important socioeconomic sector that has gained prominence over the years (MTIB 2020).

Although the issue related to sustainable supply of wood resources continue to daunt the Malaysian furniture industry, the use of other materials, such as bamboo, rattan, plastic, metal, wood-based panels, leather, etc., appears limited due to cost competitiveness. This is particularly true in the case of China and Vietnam, which enjoys a large price competitiveness in the other types of materials due to economies of scale (IFRG 2019). Currently, Malaysia is the third largest producer and exporter of furniture in Asia, after China and Vietnam. In spite, of the competition regionally,

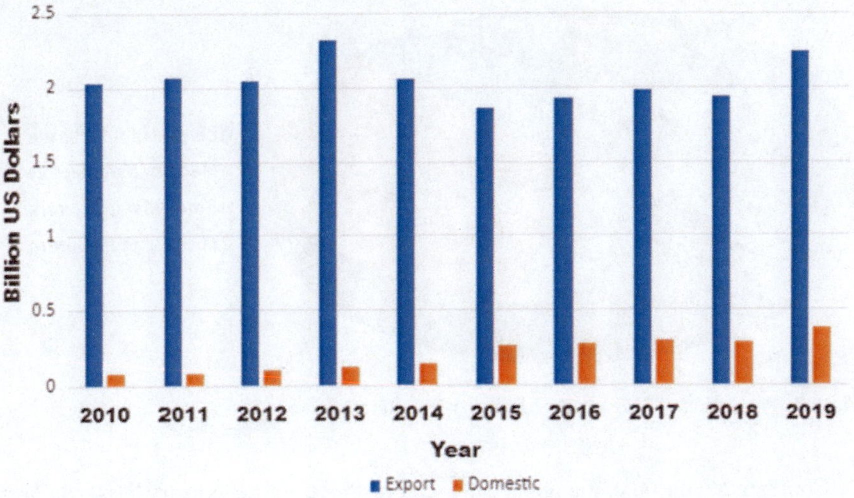

Fig. 1.2 Export and Domestic Market for Malaysian Wooden Furniture

the Malaysian wooden furniture industry has remained strong in terms of exports and the domestic market (Fig. 1.2).

1.5 Small and Medium Enterprises in the Furniture Industry

The furniture manufacturing industry has emerged as the most important sector within the wood-based industry in Malaysia, contributing nearly 1.5% to the country's GDP (MTIB 2020). From the 3500 registered furniture manufacturers in the country, 85% are classified as small and medium enterprises (SMEs). According to the SME Corporation of Malaysia, SMEs in the manufacturing sector are defined as firms with annual sales turnover not exceeding USD12.5 million, or with the number of full-time employees not exceeding 200.

Nevertheless, it must be recognized that the definition of SMEs varies across countries. According to the ITC (2018), middle-income and high-income countries consider enterprises employing up to 249 persons as SMEs. Within this definition, it is further broken down into micro-enterprises (1–9 employees), small (10–49 employees), and medium (50–249 employees). Lower income economies on the other hand, refer to enterprises using 50 or 100 employees as a threshold for SMEs. Annual turnover threshold for SMEs ranging from USD50 million to USD70 million are applicable to high-income economies, while lower income developing countries take the turnover value of USD1 million to USD5 million to define SMEs.

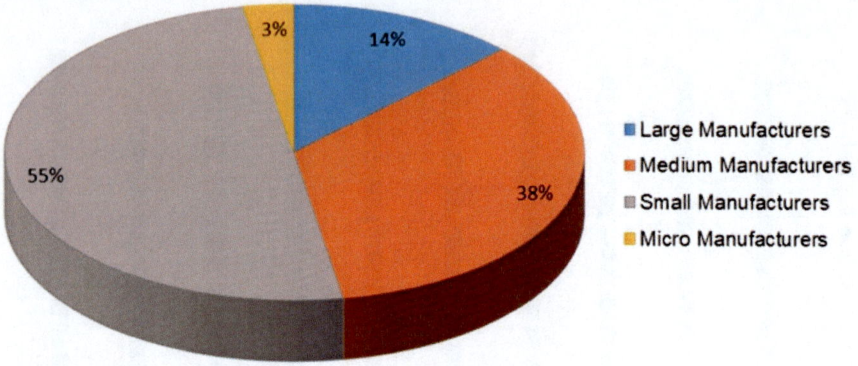

Fig. 1.3 Proportion of SMEs in the Malaysian Furniture Industry

Compared to the other manufacturing sectors, the furniture industry has the largest proportion of registered SMEs (SME Corp. 2019). This is to be expected as the furniture industry is predominated by self-made entrepreneurs, who started their business as a cottage industry, catering for the domestic market. A good example of this is a Muar furniture cluster, which started off as sub-contractors to the many Taiwanese export-oriented furniture exporters in the vicinity. Figure 1.3 shows the predominance of the SMEs in the Malaysian furniture industry.

The success of the furniture industry in Malaysia and the role of the furniture clusters located throughout the country has been well researched (Ratnasingam 2015). It is also worth noting that the furniture clusters in the locality of Muar, Batu Pahat, Melaka, Sungei Buloh, Taiping, Kuantan, and Sungei Petani, which are spread throughout Malaysia, contribute nearly 85% of the total furniture produced in the country (MFC 2017). These industrial clusters are well organized and self-sustaining and has an extensive network of supporting industries which not only provides all the necessary materials, supplies, services, logistics, shipping, etc., which explain the very short delivery time required (Ratnasingam 2015). With the large numbers of sub-contractors, specializing in parts, components manufacturing, and sub-assemblies, these furniture clusters offer both design diversity and production volume flexibility.

Categorically, SMEs in the furniture industry strive on the short manufacturing lead time, which is highly desirable to customers. This characteristically lean and mean manufacturing system accounts for the relatively competitive pricing of their products. At the national level, it has been reported that the furniture industry, particularly the SMEs, contributes a significant amount towards the national income. According to the report by MFC (2019), the total direct revenue, including taxes, contributed by the SMEs in the furniture industry amounted to USD0.5 billion in 2018, and in total the socioeconomic contribution of the furniture manufacturing industry was estimated to stand at USD17.8 billion. Further, as an important player in the country's fashion sector, and in light, of its highest added value among the wood product sub-sectors, the furniture manufacturing industry remains relevant in the overall economic pie of the country. As the furniture industry embarks towards

Fig. 1.4 Challenges of SMEs in the Furniture Sector

transforming itself into the manufacture of higher value-added furniture, with brand recognition, the sustained growth of the furniture industry will remain promising for many years to come (MFC 2019).

Despite their agility and success in furniture manufacturing, SMEs throughout the world are characterized by their limited production capacity, poorly invested facilities, not up to date technologies, lack of financial strength, and non-scientific management or rather family-run management system. The main constraints faced by SMEs in the furniture manufacturing industry is illustrated in Fig. 1.4, and it is based on a survey of SMEs throughout the Asian furniture industry as reported by Ratnasingam (2020).

As a result, in many of the SMEs in the furniture industry, the operations are not optimized and not as productive as it could potentially be. The lack of understanding and knowledge behind the art and science of furniture manufacturing continue to plague the furniture industry in many developing countries, with low productivity growth and stagnating value-addition (ITTO 2005). Inevitably, there is an apparent emphasis on experience and craftmanship, rather than scientific and engineering knowledge to ensure the sustainability of the furniture manufacturing industry in the globalized world.

Further, unlike other manufacturing industries, the application of production engineering concepts in the furniture industry is also limited, due to the lack of knowledge workers. Such a scenario explains the poor adoption of automation and advanced technologies in the furniture manufacturing industries, especially within the wooden and leather manufacturing segments. However, the wood-based panels, metal, and plastic furniture manufacturing are more akin to the application of advanced technologies, which results in higher productivity levels (Ratnasingam 2020). For example, the German furniture industry which specializes in flat-packed furniture production using wood-based panels, records very high productivity levels due to the widespread application of automation and advanced technologies in its manufacturing facilities.

Against this background, ensuring knowledge transfer to the SMEs to transform them into high productivity manufacturing enterprises continue to be the primary goals of many furniture promotion agencies throughout the world (ITTO 2005). Although the entry barrier into the industry at the craftsman level is relatively low, in many instances, these furniture manufacturing enterprises get trapped into mass production activities driven primarily by incremental inputs rather than productivity gains. Under such circumstances, the furniture manufacturing industry competes on cost competitiveness rather than value addition, which cannot be sustained in the long term. As described in the report by Ratnasingam (2015), competing based on price competitiveness is not a recipe for growth, and its adverse effects may severely impact the long-term economic viability of the business.

Furniture clusters in countries such as Italy, Taiwan, Vietnam, China, Poland, the USA, and Denmark, are predominated by SMEs, which are well researched, and its economic successes acknowledged (Hongqiang et al. 2012; Ratnasingam 2015). It has been reported that SMEs make up at least 60% of all furniture manufacturing establishments in the developed countries, while in developing countries the proportions of SMEs operating in the furniture industry is higher at 75% (ITC 2018). In this context, the predominance of SMEs in furniture manufacturing industries throughout the world is apparent, and their importance cannot be underplayed. Therefore, understanding the characteristics of SMEs, and providing knowledge on the art and science of furniture manufacturing, will contribute towards improving the overall productivity in furniture manufacturing industries, in a highly globalized and competitive world.

Summary

- Furniture manufacturing is a global industry.
- It is predominated by small and medium enterprises (SMEs).
- Furniture is fashion, and its form and function has evolved over time.

References

Bullard SH, West CD (2002) Furniture manufacturing and marketing: eight strategic issues for the 21st century. Forest and Wildlife Research Center, Bulletin FP 227, Mississippi State University, Starkville, USA

CSIL (2019) World furniture outlook. CSIL—Centre for Industrial Studies, Milan, Italy

Drayse MH (2011) Globalization and innovation in a mature industry: furniture manufacturing in Canada. Reg Stud 45(3):299–318

Hongqiang Y, Ji C, Nie Y et al (2012) China's wood furniture manufacturing industry: industrial cluster and export competitiveness. Forest Prod J 62(3):214–221

IFRG (2019) Analysis of the Asian furniture industry. International Furniture Research Group Report No. 14, Singapore

ITC (2018) Small and medium enterprises (SMEs) in the resource-based industries—an analysis. International trade Centre Technical Report No. 11, Geneva, Switzerland

ITTO (2005) International wooden furniture markets—a review. International Tropical Timber Organization Publication, Yokohama, Japan

MFC (2017) Annual report on the furniture industry in Malaysia. Malaysian Furniture Council, Kuala Lumpur, Malaysia

MFC (2019) Annual report on the Malaysian furniture industry. Malaysian Furniture Council, Kuala Lumpur, Malaysia

NATIP (2009) National timber industry policy. Ministry of Plantation Industries of Malaysia, Putrajaya, Malaysia

MTIB (2020) Annual report on the Malaysian wood industry. Malaysian Timber Industry Board, Kuala Lumpur, Malaysia

Quesada HJ, Gazo R (2006) Mass layoffs and plant closures in the US wood products and furniture manufacturing industries. Forest Prod J 56(10):101

Ratnasingam J (2015) The Malaysian furniture industry—unravelling Its growth and challenges to innovation. Universiti Putra Malaysia Press, Serdang, Malaysia

Ratnasingam J, Ioras F (2003) The sustainability of the Asian wooden furniture industry. Holz Roh Werkst 61(3):233–237

Ratnasingam J, Ioras F (2005) The Asian furniture industry: Reality behind the statistics. Holz Roh Werkst 63(1):64–67

Ratnasingam J, Ark CK, Mohamed S et al (2017) An analysis of labor and capital productivity in the Malaysian timber sector. BioResources 12(1):1430–1446

Ratnasingam J, Chin KA, Abdul Latib H et al (2018) Innovation in the Malaysian furniture industry: drivers and challenges. BioResources 13(3):5254–5270

Ratnasingam J, Khoo A, Chee JN et al (2020) How are small and medium enterprises in Malaysia's furniture industry coping with COVID-19 pandemic? Early evidences from a survey and recommendations for policymakers. BioResources 15(3):5951–5964

Schuler A, Buehlmann U (2003) Identifying future competitive business strategies for the U.S. furniture industry: benchmarking and paradigm shifts. Northeastern Research Station Technical Report, USDA-Forest Service, Newton Square, Pennsylvania, USA

SME Corp. (2019) Status of the SMEs in the manufacturing sector. Small & Medium Enterprises Corporation of Malaysia, Petaling Jaya, Malaysia

Chapter 2
Furniture Design and Innovation

Furniture design has evolved over human civilization. It's a reflection of lifestyle, material, and technology of the time. The study of furniture design involves human civilization, history of art and craft, cultural, social, and economic aspects. In this respect, furniture design is an art as well as a science, where design concepts are transformed into products using materials and processing technologies, to provide the customer a product of their preference. Inevitably, furniture design is the basis on which value of furniture is evaluated, and thus design and innovations are important to move up along the product value-chain. This chapter will discuss the salient aspects involved in the furniture design process, the evolution in design, and the sources of innovations.

2.1 Introduction

Furniture design is an integral part of human civilization. The history of furniture design has been shaped by every spectrum of the society, including royalties, artisans, architects, designers, and even the layman (Postell 2007). Early historical records have shown that tree stumps and rocks were used in early human civilization as furniture. The earliest records of furniture made from wood, stones and bones were discovered about 30,000 years ago by archeologists (Blakemore et al. 2006). Nevertheless, it was only during the period of the ancient Egyptians that construction techniques such as joinery evolved, to produce stools and tables.

During the ancient Greece and Roman period, the evolution of furniture design picked up pace to keep up with the needs of the palaces and forts. Furniture made of Oak wood, bulky, and heavy in ornaments were typical of the Middle Ages. The Italian Renaissance in the fourteenth and fifteenth centuries expanded furniture design even further. The European impact on furniture design came about in the seventeenth century, especially with the rise of Baroque designs. The nineteenth century revived styles, and the transformation towards modernism began as the new

J. Ratnasingam, *Furniture Manufacturing*, Design Science and Innovation,
https://doi.org/10.1007/978-981-16-9412-7_2

millennium approached. It must be emphasized that the unique characteristics of post-modern furniture design is a return to natural shapes and textures.

Furniture design reveals the thinking about and the making of objects built-ins, encompassing evolving techniques, materials, and available resources that are shaped by cultural, political, and economic factors. Therefore, the history of furniture design is a chronology of how societies have thought about design, fabrication, using an enormous amount of anthology of furniture (Postell 2007). On this account, any study of history of furniture design should review the historical path taken in the evolution of furniture design till the present day. In this context, an understanding of furniture design history would necessarily require a study of human civilization, lifestyle, and cultural aspects. For an elaborate discussion on the history of furniture design, the readers are referred to Blakemore et al. (2006), Postell (2007), Smardzewski (2015).

2.2 Evolution in Furniture Design

Furniture design is as much an art as a science. It continuously evolves through time, encompassing innovations, to produce furniture that meet the expectations of customers from various cultural, belief system, and lifestyle backgrounds. Furniture has deep architectural roots, reinforcing the concept of place and completing interior space, but it is also tethered to craft and industrial production (Postell 2007). Driven by a combination of structural, utilitarian, and esthetic concerns, furniture reveals a chronology of ideas beginning with the earliest societies that lived in 1700s to the present twenty-first century. A quick review of the design evolution throughout the centuries is shown in Fig. 2.1.

- **Neolithic Period/The Classical World** (3,000 B.C.–eighth century), provided the world an insight into the early human civilization where furniture was made from stone and wood for the sole purpose of daily utilities, such as storage and they were often adorned with gold, silver, ivory, and ebony for decoration purposes.
- **Early Modern Europe/Medieval Period** (500–1500 A.D.), produced furniture that were artistic, made from Oak wood and heavily ornamented. It was during this furniture design evolution accelerated.
- **Nineteenth Century** (1801–1900), made furniture designs detailed and artistic, as defined by *Gothic*, Neoclassicism, Rococo, Arts and Crafts Movement and the Art Nouveau *Movement*. With fine workmanship, handcraft, and natural designs, the furniture was mostly sought after by the wealthy and aristocratic communities.
- **Early North American** (twentieth century), produced furniture with more utilitarian roles, rather than its esthetic appeal. Simplified designs were preferred, with greater emphasis on form and function.
- **Modernism** (Post-World War II), was born out of fusion between many different styles and movements of the early twentieth century, including Bauhaus, Art Deco, and Futurism occurred. Furniture production embraced the aspects of functionality, elegance, and ultra-modern designs.

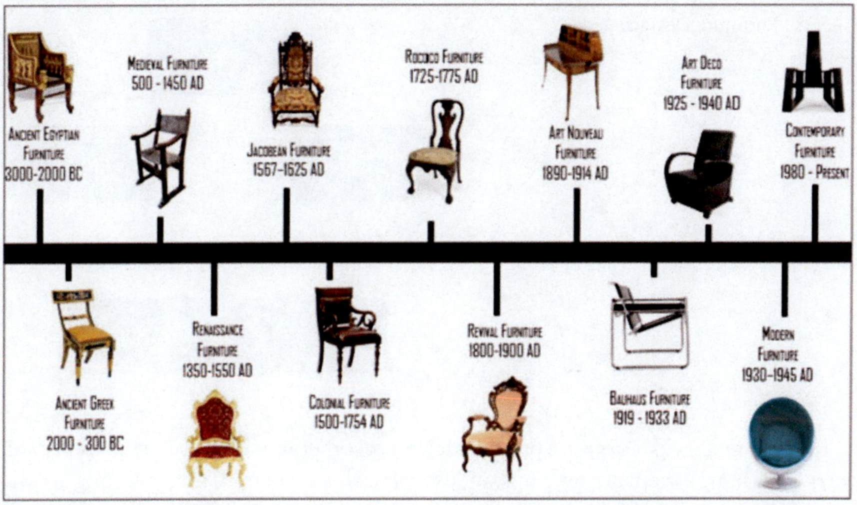

Fig. 2.1 Chronology of Furniture Design Evolution (Adapted from Csanády et al. 2019)

- **Eco-design** (1920s onwards), the concept of environmental sustainability living, which aims to strike a balance between humans and nature is the core theme of such furniture. It has gained traction, especially among generation-X and millennial consumers, who are more environmental conscious.
- **Contemporary** (1970s onwards), furniture is for the transformational lifestyle of humans of the present generation. With advancements in science and technology, furniture with simple designs, minimalistic yet functional are typical of furniture in this present day. The demand for value for money furniture, typical of Scandinavian designs are examples of such furniture.

2.3 Furniture Design Process

Furniture design is the activity of ideating a new product, that can be manufactured and sold to customers. Ideas generated through a market-sensitive thinking process often leads to new furniture product development (Fig. 2.2). In a systematic approach, furniture designers conceptualize and *evaluate* ideas, turning them into tangible inventions and products. Therefore, the furniture designer's role is to combine art, science, and technology to create new furniture products that people can use (Vickery et al. 1997).Their evolving role has been facilitated with the advances in *digital* tools and technology that now allow furniture designers to do things that include communicating, visualizing, analyzing, *3D-modelling, 3D-printing, rapid prototyping,* and carving out big data to predict consumer behavior. Such advancements have enabled the lead time from design ideas to tangible prototypes to be significantly reduced (Xiong et al. 2017).

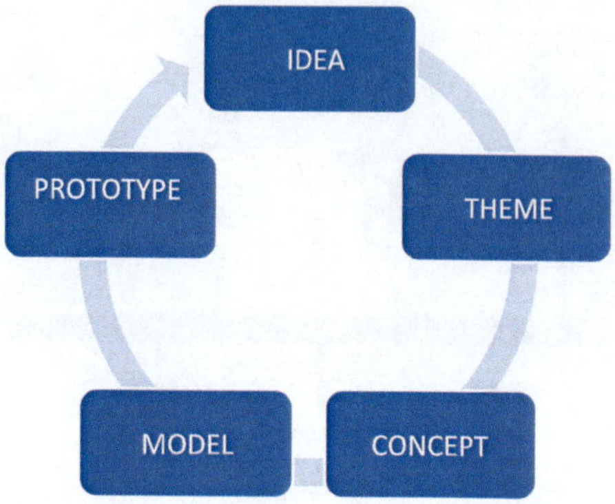

Fig. 2.2 Furniture Design Ideation Process

In some instances, furniture product design is sometimes confused with *industrial design*. Although there are overlaps, industrial design is focused on exploiting artistic form and usability, related to craft design and *ergonomics*, while furniture design is embedded closely with different human lifestyles (Fig. 2.2). Therefore, the furniture designer must appreciate the lifestyle concept in which his/her product will be used.

In general, the furniture design process typically involves three main aspects, i.e., analysis of trends, followed by the design conceptualization, and finally the synthesis of the design theme and motive (Ratnasingam 2003; Trigkas et al. 2012). In practice however, the latter two aspects are most often revisited (e.g., depending on the revisions required based on feedback from customers, buyers, or even retailers). However, this cycle of continuous feedback must also be from the designer, manufacturer, and retailer to ensure the best furniture design is produced for the customer in an economical manner (Ratnasingam 2003).

Generally, furniture designs fall under one of two categories of innovation: (1) *demand-pull innovation* (DPI), or (2) *invention-push innovation* (IPI). The former arises from an opportunity in the market that is exploited by the new product design, and it is focused on solving a market need. For instance, the SOHO furniture during the global COVID-19 pandemic is a good example (Ratnasingam 2020). On the other hand, the latter type of furniture designs is driven by advancements in materials science, technology, or a transformational lifestyle (Fig. 2.3). This are furniture designs based on new materials, processes, or a completely new form over function (UNCTAD 2018).

In this respect, the furniture designer needs to consider how the furniture may be abused, potential faulty design element, manufacturing errors, and most importantly the recommended use of the furniture. Previous reports have shown that many new designs do not even leave the drawing table, while others do not proceed beyond the prototype stage (Ratnasingam 2003; Postell 2007). Other designs will become obsolete and be discarded from the market. In this respect, the furniture designer may be frustrated repeatedly as success rate is usually in the range of 10–15% with a new product design (Binz and Truffer 2017). Listening to the market needs is crucial in successful furniture design.

Fig. 2.3 Lifestyle Concepts (Courtesy of Malaysian Furniture Council)

It must however be recognized that all furniture designs are clearly a source of competitiveness to the furniture manufacturing industry. Innovations is the foundation for the new products development. Innovations in furniture can be brought about by innovative new materials, advanced processing technology, and of course new design themes and interpretations. Figure 2.4 shows the various steps involved in the furniture design process. As highlighted in the report by Ratnasingam et al. (2018), the extent of innovations within the furniture industry is relatively limited and most sources of innovations are external. Nevertheless, it must be emphasized that without innovation, the opportunity to move along further the value-chain to

Fig. 2.4 Furniture Design Process

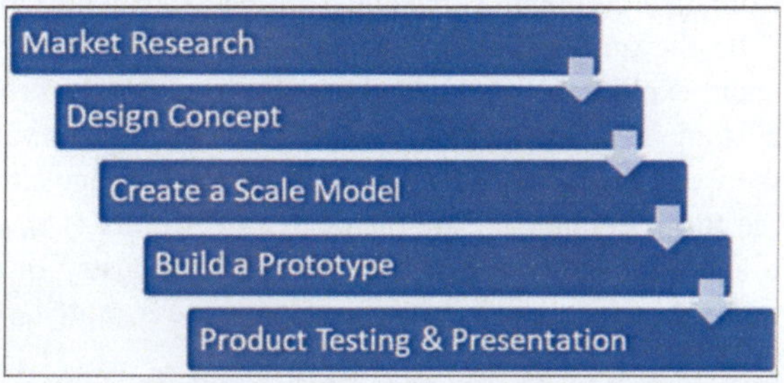

starve off competition is limited. In essence furniture is designed to benefit the full spectrum of people, accommodating their needs both at home and workplace.

2.4 Relationship Between Innovation and Design

Innovation relates to new product development, technological advancement, manufacturing scheme, discovery of materials, and marketing models. In essence, innovation involves originality and improved effectiveness, that leads to better performance. However, innovation cannot be construed as being invention. In simple term, innovation is best described as the practical implementation of an invention to have an impact in the market, but not all innovations require new inventions (UNCTAD 2018). Technical innovations are brought about by improved engineering processes, while in economics and management sciences, innovations are usually related to new models that affect the market or society. In the sphere of industrial economics, improvements in livelihood are often attributed to innovations in services or novel ways of doing business (UNCTAD 2018; Ratnasingam et al. 2018).

The fact that furniture is fashion, and it is continuously evolving to suit the customer needs implies that furniture design and innovation are intertwined. This point must be articulated since the extent of value-addition in furniture is often underpinned by the degree of innovation the product has undergone, which enables it to move along further the value-chain. Therefore, innovative furniture products often reflect new designs, but may also manifest as the use of new materials or manufacturing processes.

2.5 Sources of Innovation

Generally, innovation is a systematic approach by companies to undertake one or several activities to make them more competitive in the marketplace. It may revolve around new product design, application of advanced technologies, improved manufacturing processes, or simply exploring into a new market, which differentiates them from their competitor (Ng and Thiruchelvam 2011, 2012).

In the furniture industry, innovation involve both processes and outcomes. It is about exploiting value-added novelty in materials, processes, designs, markets, and customers' social spheres to create greater perceived value, thus moving along further the value-chain. The two main dimensions of innovation that is observed closely in the furniture industry are the degree of novelty (i.e., whether an innovation is new to the firm, new to the market, new to the industry, or new to the world) and kind of innovation (i.e., whether it is product, process, or material innovation) (Thiruchelvam et al. 2013).

The fact that furniture is fashion suggest that it is closely intertwined with human lifestyle, spatial and behavioral processes. Understanding lifestyles often generates novel ideas that could be transformed into innovation, a cornerstone that is often overlooked by furniture manufacturers. Furniture being merchandise cannot be treated in similar perspectives as other wood products, which are commodity-type products. Failure to observe this difference remain the biggest hindrance to enable greater value addition in the furniture industry (Ratnasingam 2015).

The fact that furniture manufacturing is an industry characterized by its low-entry barrier, low investment environment, and predominated by small and medium enterprises (SMEs), it is no surprise that the sources of innovation in the industry is often from third parties (Ratnasingam 2015; Ratnasingam et al. 2018). Suppliers, furniture buyers, and machinery support personnel are among the leading provider of innovative information to furniture manufacturers. Inevitably, it is well established that investments into research and development (R&D) activities in the furniture industry throughout the world is limited, and most innovations are driven by external sources (UNCTAD 2018).

2.6 Environmentally Sustainable Furniture Design

Product design, built environment, and services that embraces the philosophy of ecological sustainability is often known as eco-design or environmentally sustainable design (Ratnasingam and Wagner 2009). The primary objective of sustainable design is to minimize if not eliminate completely the negative impact on the environment, by reducing waste, skillfully working around sensitive designs, and sustainable consumption of resources. In fact, manifestations of sustainable design require *renewable resources* and innovations to have minimal impact on the environment, while connecting people and environment in harmony.

It has been highlighted that the world's greatest challenge is not environmental pollution, but rather poor product design. On hindsight, most pollutions are caused by products that involved wasteful manufacturing processes, emission of pollutants, and over extraction of resources leading to scarcity (Xiong et al. 2020). In fact, it has been reported that good designs promote extended product service life and living in pleasure and abundance with nature.

Design-related decisions are made so frequently that it has a profound impact on realizations of the Sustainable Development Goals (SDGs). Humans need to appreciate the fact that *sustainability* and design are intimately linked, and in this context the future of mankind is designed. Therefore, what is required is for design practitioners, who conform to making of products, services, as well as business which are sustainable with a positive carbon footprint.

Summary

- Innovation is essential in the furniture manufacturing industry to ensure the furniture manufacturer moves along the value-chain.
- The main sources of innovation in the furniture industry are external sources, rather than in-house R&D activities.
- Furniture design must be both creative, innovative yet production friendly to ensure its success in the marketplace.

References

Binz C, Truffer B (2017) Global Innovation Systems—a conceptual framework for innovation dynamics in transnational contexts. Res Policy 46(7):1284–1298

Blakemore RG (2006) History of interior design and furniture—from ancient Egypt to 19th Century Europe. Wiley, Hoboken, New Jersey

Csanády E, Kovács Z, Magoss E et al (2019) Optimum design and manufacture of wood products. Springer International Publishing, Cham, Switzerland

Ng BK, Thiruchelvam K (2011) Sectoral innovation systems in low-tech manufacturing manufacturing—types, sources, drivers, and barriers of innovation in Malaysia's wooden furniture industry. Int J Inst Econ 3(3):549–574

Ng BK, Thiruchelvam K (2012) The dynamics of innovation in the Malaysian wooden furniture industry—innovation actors and linkages. Forest Policy Econ 14(1):107–118

Postell J (2007) Furniture design. Wiley, Hoboken, New Jersey, USA

Ratnasingam J (2003) A matter of design in the Southeast Asian wooden furniture industry. Holz Roh Werkst 61(2):151–154

Ratnasingam J (2015) The Malaysian furniture industry—unravelling its growth and challenges to innovation. Universiti Putra Malaysia Press, Serdang, Malaysia

Ratnasingam J (2020) SOHO Furniture—a new trend in the making. Tech. Note No. 3. IFRG Publication, Singapore

Ratnasingam J, Wagner K (2009) Green manufacturing practices among wood furniture manufacturers in Malaysia. Eur J Wood Wood Prod 67(40):485–486

Ratnasingam J, Chin KA, Latib HA et al (2018) Innovation in the Malaysian furniture industry: drivers and challenges. BioResources 13(3):5254–5270

Smardzewski J (2015) Furniture design. Springer International Publishing, Cham, Switzerland

Thiruchelvam K, Chandran VGR, Ng BK et al (2013) Malaysia's quest for innovation—progress and lessons learnt. SIRD Publication, Petaling Jaya, Malaysia

Trigkas M, Papadopoulos I, Karagouni G (2012) Economic efficiency of wood and furniture innovation system. Eur J Innov Manag 15(2):150–176

UNCTAD (2018) Technology and innovation report. United Nations conference on trade and development, Vienna, Austria

Vickery SK, Dröge C, Markland RE (1997) Dimensions of manufacturing strength in the furniture industry. J Oper Manag 15(4):317–330

Xiong X, Ma Q, Wu Z et al (2020) Current situation and key manufacturing considerations of green furniture in China: A review. J Clean Prod 267:121957

Xiong XQ, Guo WJ, Fang L et al (2017) Current state and development trend of Chinese furniture industry. J Wood Sci 63:433–444

Chapter 3
Furniture Manufacturing Systems

The furniture manufacturing system has evolved from an artisan, skilled-based system, to one that employs high-technology machines. Such a transformation of manufacturing system has enabled significant improvements in production rates, and a wide range of products to be manufactured with precision and quality. The application of computer numerical technology and other allied computer-based systems has brought notable changes to an industry that has traditionally been dependent on human skills to produce furniture. This chapter will provide an overview of the transformation that has taken place in furniture manufacturing systems.

3.1 Overview of Manufacturing Systems

Generally, furniture encompasses both decorative art and design. Apart from a utilitarian role, furniture has also played symbolic and religious roles in human civilization. Made from many different types of materials, its construction has evolved from using primitive tools to constructions based on woodworking joints that reflected the local culture as well as the artisan's skills. In the early days, furniture manufacturing has been time consuming, as the craftsmen turn ideas into physical products using rudimentary tools and equipment (Ratnasingam 2015).

Since the industrial revolution, with the discovery of electric power and mechanical wood processing technology, furniture manufacturing evolved to keep pace with the developments. Power tools and machines were incorporated into the manufacturing processes, and the craftsman's skills were overtaken by the accuracy and speed of technology. Although many of the woodworking machine technology was adopted from the principles of metal working machines, nevertheless, it's adoption completely revolutionized the furniture and woodworking industry (Ratnasingam and Tanaka 2002; Black and Kohser 2019).

J. Ratnasingam, *Furniture Manufacturing*, Design Science and Innovation,
https://doi.org/10.1007/978-981-16-9412-7_3

The advent of machine woodworking resulted in improved design and manufacturing productivity in the furniture and wood-based industry. Two important characteristics of machine woodworking application is the increased machining accuracy and faster production rates, which until then, could not be achieved by the skilled craftsmen, especially for repetitive operations (Ratnasingam et al. 1999).

The introduction of information and computer technologies (ICT) to furniture and woodworking industries revolutionized the industry further. Computer-aided design (CAD) and computer-numerical controlled (CNC) machines were among the early technologies adopted by the furniture industry worldwide (Simek and Sebera 2010). These technologies used in combination enabled manufacturers to boost their design and manufacturing processes to levels of speed and accuracy not achieved before.

CAD software is a powerful tool that enables the designer to visualize, improvise, and alter designs quickly on the computer screen. It also allows the depiction of 3D designs of furniture, which improves the design ideation and product development significantly (Zhao and Li 2014). The enhanced clarity of design ideas, and the ability to rotate, alter, and render colors quickly, means the product development lead time is significantly reduced. In fact, the CAD software improves creativity, precision, speed, and accuracy in product development, while reducing cost.

The interphase between computers and machines through computer languages, such as the G-codes and M-codes, resulted in computer-controlled machines. For instance, the CNC routers guided by CAD data executes machining operations with high precision and quality of finish. In fact, CNC workstations have replaced humans in many repetitive operations in many industries and allows a variety of machining operations to be carried out with quick automated tool changes. The CNC router easily machines components and decorative parts for furniture once the desired shapes and dimensional data are programed into the computer (Gawroński 2013).

In general, furniture manufacturing systems has also evolved with time. The traditional craftsmanship has been replaced by technology to facilitate mass production systems (Yao and Carlson 2003). Mass production systems was facilitated by machine technology, which allowed product manufacturing in large quantities, with minimal diversity. Its manufacturing flow is continuous, where at every workstation an operation on the product is carried out, as it moves through the line until finish (Fig. 3.1).

As the demand for product diversity increased, especially in fashion-sensitive items such as furniture, batch production system evolved. Such a system is characterized by small production quantities, with larger product diversity. With Industry 4.0 making its entry almost a decade ago, batch size 1 (BS1) production system was introduced in the manufacturing sector (Muhammadi et al. 2020). Batch size 1 production is a completely different model from the traditional mass production model. The smaller the batches go, the more flexible and intelligent the machines need to be, and therefore manufacturers need to change the equipment, the method, and the planning of furniture production.

Despite the technological advancements available in the marketplace, the one-piece production system is sometimes employed to suit individual customer's requirement for unique furniture pieces. For instance, in replicating antique furniture using

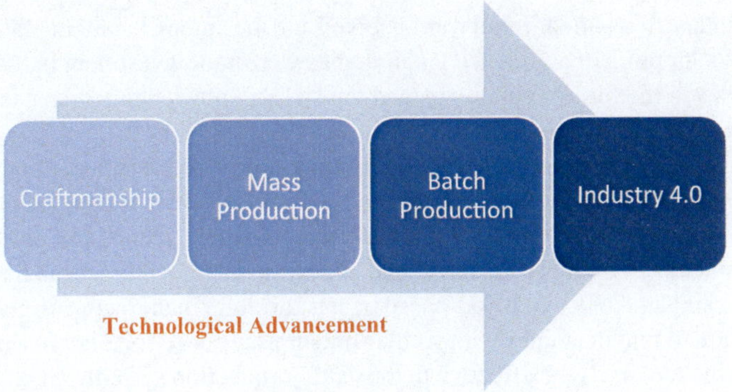

Fig. 3.1 Evolution of the furniture production system

exceptional raw materials, to produce one-of-a-kind replica of the highest quality, craftsmanship, and esthetic appeal, the work of the master craftsman is indispensable (Yao and Carlson 2003; Hunter 2008). Such a production system is however rare due to the lack of such master craftsmen, and the hefty costs involved.

3.2 Application of Technology in Furniture Manufacturing

Technology and automation have often been advocated as the long-term solution for the global labor-intensive furniture manufacturing industry, especially in the developing world (Das 1996). This argument has been put forward on the premise that labour productivity has been stagnating in the furniture industry and is relatively lower compared to other resource-based manufacturing sectors. Inevitably, the labor cost per unit production is comparatively higher in furniture manufacturing, which could be significantly reduced through the adoption of technology and automation (Ratnasingam et al. 1999). The Academy of Sciences of Malaysia reported that the uptake of technology and automation solutions among furniture manufacturers were relatively low, and even those large-sized establishments preferred low-cost automation solutions and stand-alone CNC workstations capable of performing repetitive tasks at higher production rates (ASM 2017).

Generally, automation in manufacturing sectors fall under one of the three categories; (i) low-cost automation (LCA), (ii) mid-range automation, which requires skilled supervision to ensure conformity, and (iii) high-range automation, which is devoid of any supervision, and operates automatically through a machine-human interphase (Ratnasingam et al. 2018). In a survey of the furniture manufacturers in Southeast Asia, it was found that almost half of all automation technologies used fall under category 1, while categories 2 and 3 accounted for 38% and 12%, respectively, of the automation technologies used (Ratnasingam 2019). In a recent study by Ratnasingam et al. (2018) on the extent of automated technologies application in the furniture industry, a similar trend was observed and the among the main reasons cited for the poor adoption of automated technologies were poor investment in ICT-related technologies in the furniture industry, and the lack of knowledge workers capable of supporting such technologies.

Nevertheless, the readiness among furniture manufacturers to adopt automation and advanced technologies is limited among wooden, rattan, bamboo, and leather furniture manufacturers, while wood-based panels, metal, and plastic furniture manufacturers are more receptive and appear to readily embrace such technologies (Ratnasingam 2019). In wood-based panels furniture manufacturing companies, it is common to find machines with similar functions grouped together in a particular section in the factory, to be effective in the batch-production systems. Such workstations often have several CNC machines, which could change tools automatically to perform operations, such as routing, sawing, boring, and shaping, without having to move to another workstation (Fig. 3.2). In essence, the application of such technologies offers significant economic benefits for the furniture manufacturer, and it enables a quick changeover of design as well as production volume. Work study carried out in furniture factories have shown that such automated technologies promote a balanced workload, while streamlining the production processes to optimize capacity utilization, to ensure shorter manufacturing lead time (Ratnasingam 2016).

In furniture manufacturing, CNC workstations have transformed the industry significantly. Not only different machining operations can be carried out fast and accurately but also tool change time can be significantly reduced as different tools kept in a tool magazine in the machine allows quick tool changes. Through variable cutting speeds for tools of different diameters, machining quality is also improved. In fact, the application of CNC workstations in furniture manufacturing have been reported to reduce handling time, tool change time, and machining time by 30%, 18%, and 11%, respectively (Ratnasingam 2016). The application of robotic arms also facilitates the loading and unloading of workpieces from the CNC workstation, which further improves productivity.

In a typical furniture manufacturing operation, manufacturing lead time is highly dependent on handling, tool change, and machine downtimes. Although machining time is often fixed based on cutting and feed speeds, the other time elements are usually beyond the control of the operator (Ratnasingam and Tanaka 2002). However, with CNC workstations, these challenges have been alleviated through automated technologies. In fact, the application of CNC machines and workstations render flexible automation in furniture manufacturing, which allows quick changeover to different operations, components, and parts.

In the present-day furniture industry, manufacturing flexibility is an essential requirement for several reasons (Ratnasingam 2015).

1. New product launches take place frequently, and therefore, the flexibility to manufacture diverse products is important.
2. Producing different products will enable optimization of manufacturing capacities.

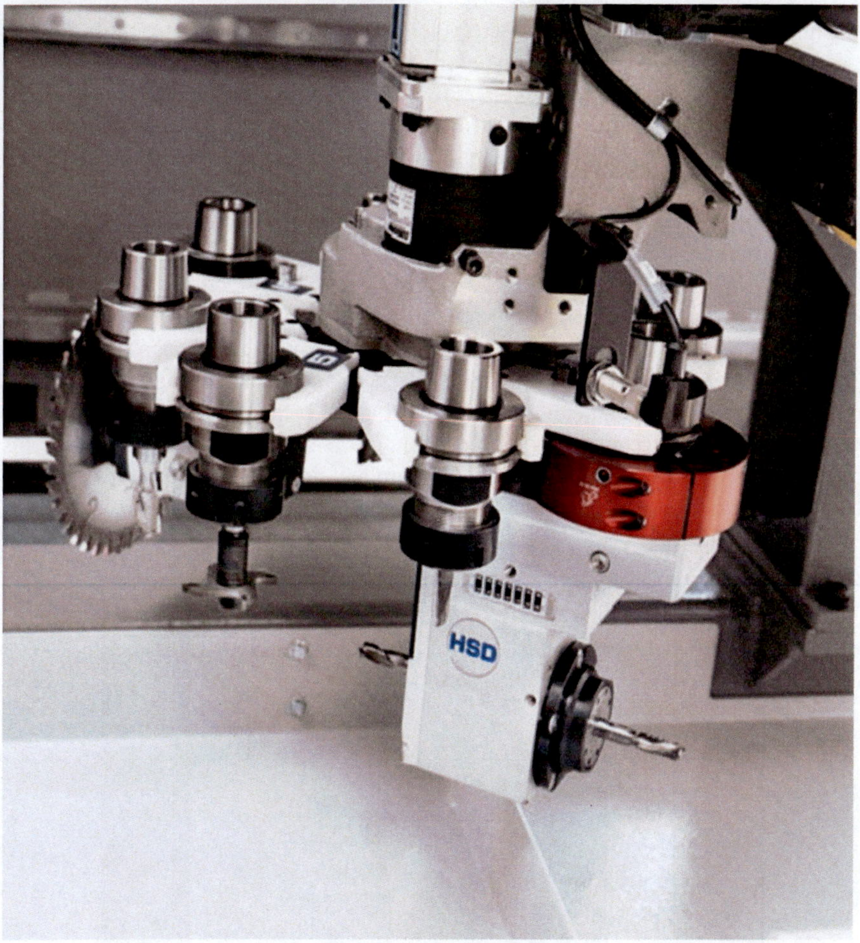

Fig. 3.2 Different cutters in a CNC machine (Courtesy of the Malaysian Furniture Council)

3. Flexibility in manufacturing paves the way for the adoption of batch production system.
4. Flexible manufacturing systems allows the furniture manufacturers to respond promptly to customers' needs.
5. Flexible manufacturing systems tend to reduce losses due machine downtime, as alternative machining workstations are readily available.
6. Flexible manufacturing systems will also specialization as well as economies of scale in furniture manufacturing.

However, flexible manufacturing systems using CNC machines and workstations pose several challenges. Errors in software, programing language, and controls are the main problems faced on the factory shopfloor. This is particularly true when new designs with intricate shapes are being machined, or when a new material is being machined and its tool wearing characteristics is unknown. Under such circumstances, close monitoring by the machine operator is crucial at least until stability is achieved in the machining and tool change operations (Altintas and Aslan 2017).

In the larger furniture manufacturing enterprises, especially those involved in the manufacture of wood-based panels furniture, the application of the advanced computer integrated manufacturing (CIM) is increasingly apparent (Eyers and Potter 2017). In such systems, the role of the computer is extended beyond design and manufacturing, into production scheduling, quality assurance, and inventory management. Figure 3.3 shows the various functions in the furniture manufacturing enterprise that is handled by the CIM system.

The computer-aided design (CAD) system incorporates the complete design process, from detailed product drawing to strength and stability evaluation of furniture, by applying the Finite Element Method (FEM). In this regard, the furniture designer is given more time to consider ergonomic, safety and esthetic requirements of the new product. Through such detailed product analysis, the furniture designer could also gather relevant information for economical manufacturing of the product.

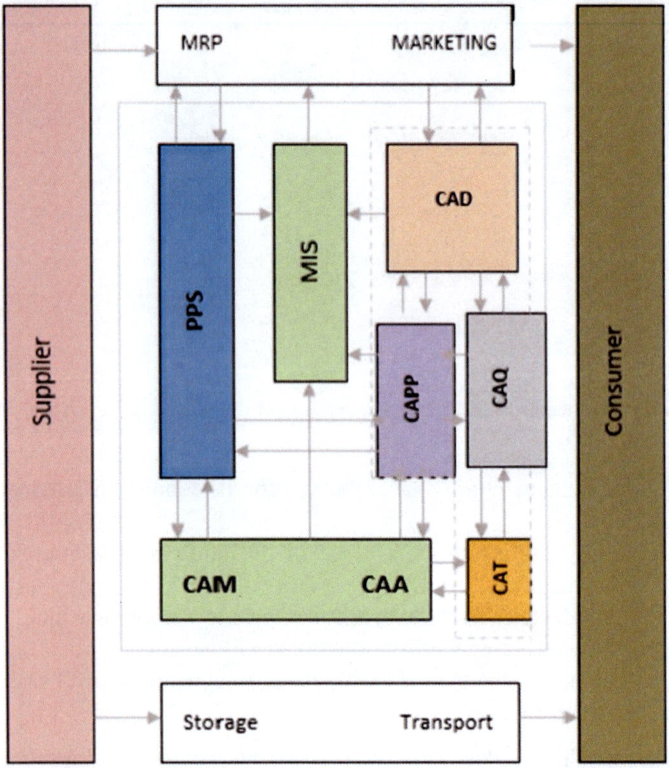

Fig. 3.3 Integrated functions in computer-integrated manufacturing system (Adapted from Csanády et al. 2019)

The CIM system incorporates a full spectrum of different computer software to undertake various functions in the manufacturing enterprise (Bortolini et al. 2018). Computer-aided engineering (CAE) comprise of both CAD and computer-aided manufacturing (CAM), computer-aided process planning (CAPP), computer-aided quality assurance (CAQ), and computer-aided testing (CAT), provides the necessary support to manage and control the manufacturing operations (Csanády et al. 2019).

In fact, computer-integrated manufacturing (CIM) comprises all activities associated with the use of computer control to manage the whole manufacturing process. Among the important activities undertaken by the computers include control and monitoring of machining operations, work transfers, movement of work-in-progress (WIP), stock transfers, costing, scheduling, work routing, and assembly instructions (Mohammadi et al. 2020).

In essence, the application of computer information technology (ICT) is inevitable in the globalized furniture industry. Such technologies are crucial to enable the furniture manufacturing industry to remain competitive and sustainable in the future.

Summary

- Furniture manufacturing system has evolved from craftsmanship to machine-based manufacturing systems.
- Mass production systems is moving towards smaller batch systems.
- The potential application of ICT in furniture manufacturing is huge and cannot be overlooked.

References

Altintas Y, Aslan D (2017) Integration of virtual and on-line machining process control and monitoring. CIRP Ann 66(1):349–352

ASM (2017) Furniture industry sector—Mega science 3.0. Academy of Science of Malaysia, Petaling Jaya

Bortolini M, Galiza FG, Mora C (2018) Reconfigurable manufacturing systems: literature review and research trend. J Manuf Syst 49:93–106

Black JT, Kohser RA (2019) DeGarmo's materials and processes in manufacturing, 13th edn. Wiley, Hoboken, New Jersey, USA

Csanády E, Kovács Z, Magoss E et al (2019) Optimum design and manufacture of wood products. Springer International Publishing, Cham, Switzerland

Das SK (1996) The measurement of flexibility in manufacturing systems. Int J Flex Manuf Sys 8:67–93

Eyers DR, Potter AT (2017) Industrial Additive Manufacturing: a manufacturing systems perspective. Comput Ind 92(93):208–218

Gawroński T (2013) Optimisation of CNC routing operations of wooden furniture parts. Int J Adv Manuf Tech 67:2259–2267

Hunter SL (2008) The Toyota production system applied to the upholstery furniture manufacturing industry. Mater Manuf Process 23(7):629–634

Mohammadi S, Mirzapour SMJ, Rekik Y (2020) An integrated production scheduling and delivery route planning with multi-purpose machines: a case study from a furniture manufacturing company. Int J Prod Econ 219:347–359

Ratnasingam J (2015) The Malaysian furniture industry – Unravelling its growth and challenges to innovation. Universiti Putra Malaysia Press, Serdang, Selangor

Ratnasingam J (2016) Work study applications in furniture manufacturing—Current status. Tech. Note No. 11. IFRG Publication, Singapore

Ratnasingam J (2019) Current manufacturing technologies in furniture industry—a review. Tech. Note No. 5. IFRG Publication, Singapore

Ratnasingam J, Ma TP, Perkins MC (1999) Productivity in wood machining processes—is it a matter of simple economics? Holz Roh Werkst 57:51–56

Ratnasingam J, Tanaka C (2002) Wood machining processes—a managerial perspective. Tanabe Foundation Publication, Shimane, Japan

Ratnasingam J, Ab Latib H, Yi LL et al (2019) Extent of automation and the readiness for industry 4.0 among Malaysian furniture manufacturers. Bioresources 14(3):7095–7110

Simek M, Sebera V (2010) Traditional furniture joinery from the point of view of advanced technologies. Paper WS-74 in proceedings of the international convention of society of wood science and technology and United Nations economic commission for Europe—timber committee, Geneva, Switzerland, October 11–14, 2010

Yao AC, Carlson JGH (2003) Agility and mixed-model furniture production. Int J Prod Econ 81(82):95–102

Zhao C, Li J (2014) Analysis and improvement of multi-product assembly systems: an application study at a furniture manufacturing plant. Int J Prod Res 52(21):6399–6413

Chapter 4
Production Flow in Furniture Manufacturing

Check for updates

The furniture manufacturing process involves a series of machining, jointing, abrasive sanding, and finishing operations, through which the rectangular sawn wood material is converted into a three-dimensional furniture product based on the design specifications. In doing so, these operations must be carried out optimally to ensure the highest quality and lowest production cost is achieved. Although furniture manufacturing has been a traditional skilled-based industry, the application of machining technology has enabled productivity improvement in the various operations. This chapter provides an overview of the various processes involved in the manufacture of furniture, of both solid wood and wood-based panels, paving the way for an in-depth discussion of these various operations in the following chapters.

4.1 Introduction

The design and manufacture of furniture is as much an art as a science. The ideation of the design is an art, while its transformation into a 3D product is a science, involving facets of material science, engineering, and construction technology (Ratnasingam 2013). Compared to the traditional furniture-making process by the master craftsmen using their hands, the present-day furniture manufacturing environment is indeed a technological marvel.

The many stages involved from the design development to the manufacture of furniture is illustrated in Fig. 4.1.

J. Ratnasingam, *Furniture Manufacturing*, Design Science and Innovation,
https://doi.org/10.1007/978-981-16-9412-7_4

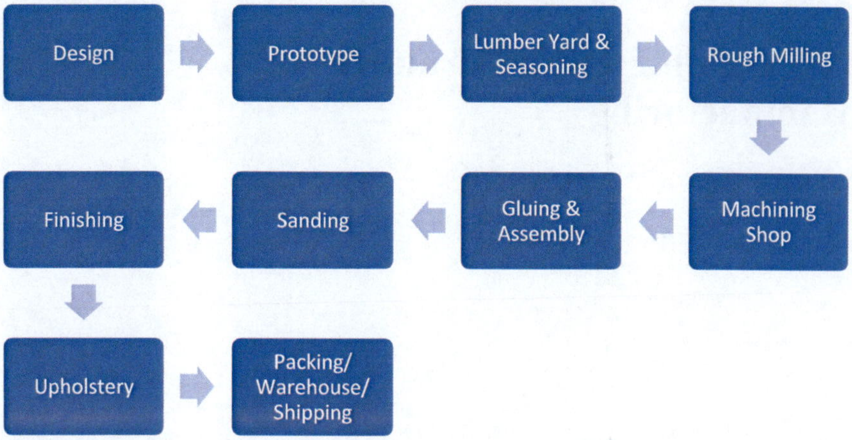

Fig. 4.1 Production flow in a furniture factory

4.2 Stages of Furniture Manufacturing

Stage 1: The ideation and conceptualization of the design is usually transformed into a prototype/sample, to be evaluated by the customer for acceptance and satisfaction (Fig. 4.2). At this stage, essential information pertaining to material requirements, production sequences, impending quality issues, packaging protocols, and product costing is gathered.

Stage 2: The material requirements for a production order is usually determined based on the Bills of Materials (BOM). The materials specifications in terms of its dimensions and quality as well as its quantity are finalized at this stage (Fig. 4.3).

Fig. 4.2 From design to prototype (Courtesy of the Malaysian Furniture Council)

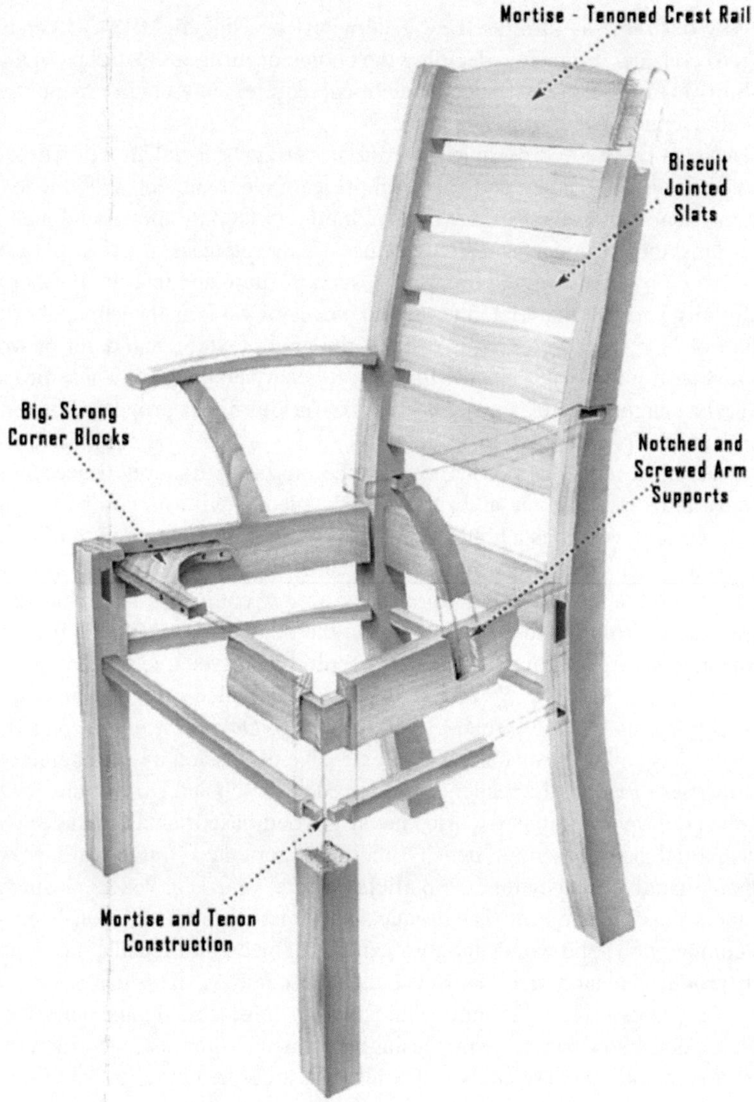

Fig. 4.3 Furniture components

Many factors are taken into consideration when deciding the choice of material for furniture. High on the priority list are the material's density and workability. Since density has a close relationship to strength, it will determine its suitability to support load, and indirectly, how easily it can be worked with. The latter is particularly important because it predetermines the processing time and cost if the material, is chosen. Further, density also determines if the material can be handled easily on the factory shopfloor. It is often highlighted that easily workable and light weight material are highly desirable as furniture raw materials (Ratnasingam 2013). Other factors taken into consideration when deciding the choice of furniture materials include its availability in the market, cost, special treatment requirements, environment-friendly, and of course the customer's acceptance.

Because wood is a hygroscopic material, it warrants special attention especially in terms of moisture content and chemical preservative treatment. Wood when used in environments with very variable relative humidity tend to 'move' and manifest a degree of instability. Under such circumstances, it also increases the risks of the wood to be attacked by biodegrading organisms, such as fungi and insects. Further wood is a naturally variable material, and no two pieces of wood is the same, although it may be from the same tree. In this respect, the grain, texture, and color of wood is variable, that it imparts the wood with natural beauty and appeal. These properties may also be sought after, by some customers of furniture. The properties of common furniture materials are shown in Table 4.1.

In general, wood is a porous and is the fibrous structural tissue found in the stems and roots of trees and other woody plants. With its complex lignocellulosic structure, wood has a high strength-to-weight ratio compared to other materials. It is a 'green' material which is sustainable. Generally, wood is classified as either hardwood or softwood. The former has a more complex anatomical structure and are usually from broad-leaved trees. On the other hand, softwood has simpler anatomical structure, and are produced by needle-like leaved tree species. The wood from conifers (e.g. *Pine*, *Fir*, *Spruce*, *Hemlock*, etc.) is called softwood, and the wood from dicotyledons (usually broad-leaved trees, e.g. *Oak*, *Meranti*, *Maple*, *Walnut, Cheery*, etc.) is called hardwood. For an extensive discussion on the characteristics and properties of wood, the readers are referred to Desch and Dinwoodie (1996).

Engineered wood composites, also known as wood-based panels, include products such as particleboard, medium density fiberboard, oriented strand board, plywood, which are manufactured by binding particles, fibers, strands or veneers, using adhesives under pressure and heat. The products have more uniform and consistent properties compared to solid wood, and they can be engineered to specifications, and are usually produced in large sizes. However, the most common size is panels of 1.2 m by 2.4 m, of thicknesses up to 25 mm. With advances in processing and adhesive technologies, wood-based panels are increasingly produced from waste wood and agricultural waste, including rye straw, wheat straw, rice straw, hemp stalks, kenaf stalks, or sugar cane bagasse. For a detailed discussion on wood-based composites, the readers are referred to Stokke et al. (2014).

Plastics are one of the most versatile material available and is made from synthetic or semi-synthetic organic compounds, which are malleable, and thus can be easily molded into solid objects. Classification of plastics are based on its properties, such as hardness, density, strength, heat, and chemical resistance. There are many types of plastics, but the common ones used in furniture are *acrylics*, *polyesters*, and polyurethanes. The most common however, polymethyl methacrylate (PMMA), an acrylic plastic, with excellent working properties and low cost, which has become the premium material for the manufacture of molded plastic chairs. Nevertheless, the success and dominance of plastics as a material has been severely marred due

Table 4.1 Properties of common furniture materials

Material	Properties
Wood 	(1) Natural and beautiful material (2) Sustainable source (3) Environment-friendly (4) Diverse physical, chemical, and mechanical properties (5) High strength-to-weight ratio (6) Durability improved through preservative treatment (7) Good working properties (8) Low carbon footprint (9) Good heat insulator (10) Can be modified through heat treatment and compression
Wood-Based Panels 	(1) Available in large sizes (2) Engineered material with uniform properties compared to solid wood (3) Produced from low-quality fiber resources, including waste wood (4) May be laminated with veneer, paper, and plastic overlays or even directly printed (5) Manufactured to specifications (6) Relatively more abrasive on the cutting tools compared to solid wood (7) Less durable and is not moisture resistant
Plastic 	(1) Polymer-based material which can withstand heat, scratching, chemicals, weather, and moisture (2) Low ductility, poor stiffness, and susceptible to creep (3) Good insulation properties (4) Relatively high carbon footprint (5) Recyclability is a global concern (6) Strong and a lightweight material (7) Can be finished easily to accommodate different patterns, profiles, and colors (8) Dimensionally stable and durable (9) Relatively low cost
Metal 	(1) Has natural lustre (2) Strong, hard, and dense, but malleable and ductile (3) Scratch and abrasion resistance (4) Lower strength to weight ratio (5) Good conductor of heat and electricity (6) High heat resistance (7) Can be combined to produce alloys (8) Susceptible to corrosion (9) Good working properties
Bamboo 	(1) Highly versatile and sustainable material (2) Lightweight and flexible (3) Strong, tough with high tensile strength (4) Relatively low cost (5) High fire resistance (6) Good working properties but tough on cutters (7) Not durable and dimensionally instable (8) Natural beauty with good texture

(continued)

Table 4.1 (continued)

Material	Properties
Rattan 	(1) High plastic deformation and flexible (2) Relatively good strength properties (3) Naturally not durable and requires preservative treatment (4) Good working properties and easily bent (5) Relatively low cost (6) Sustainable supply as it is fast growing (7) Compared to bamboo, it can be machined easily

to its slow decomposition and poor recyclability. Hence, increasingly the use of biodegradable plastics and recycling plastics are gaining worldwide popularity. A detailed discussion on the topic of plastics as a material and its processing can be found in Cybulski (2009).

Metals have traditionally been used as a furniture material for centuries, for both indoors and outdoors applications. Cast iron, which is hard, and sturdy is often used for outdoor or garden furniture manufacturing, with its rustic look. Its main drawback is however its susceptibility to corrosion in high-humidity environments. Stainless steel is increasingly used in modern home furniture manufacturing, due to its lightweight and easy workability. However, its higher cost makes it less attractive compared to aluminum. Aluminum is now the most widely used metal in the furniture industry due to its competitive cost, lightweight, corrosion-resistant, and excellent working property, especially for the manufacture of molded metal chairs. In this context, metals continue to be the preferred material for furniture making due to its contemporary look, with charm and character (Black and Kohser 2019).

Bamboos are *evergreen perennial flowering plants* in the *subfamily Bambusoideae*, belonging to the *grass* family *Poaceae*. They grow extremely fast, through two distinct patterns, i.e., clumping and running. It is a versatile resource, and has been used many Asian societies for centuries, as a food source as well as an industrial material. Bamboos have been recognized as a suitable species for afforestation activities due to its fast growth, and its ability to sequestrate carbon efficiently makes it an excellent choice as climate change mitigation Similar to wood in chemical composition, bamboos are natural composite material with very high strength-to-weight ratios. It is increasingly used for flooring and furniture production, due to attractive texture.

Rattan which encompasses 600 species are *climbing palms* belonging to subfamily *Calamoideae*. It has flexible and porous woody stems, which imparts it excellent bending and working properties. The largest rattan genus is *Calamus*, which is widely distributed throughout Asia, and the rattan specie known as *Calamus manan* is the most common. Although bamboos and rattans are important resources, they are classified under non-wood forest products (NWFP) with limited attention in the past. However, with the establishment of the International Network of Bamboo and Rattan (INBAR) in China, substantial work has been carried out on these resources.

The readers are referred to Zehui and Zhenhua (2007) for further information on these resources.

One material that is increasingly gaining a strong foothold in the furniture materials market is glass. Glass is unique and extremely versatile, and can be engineered to have specific optical, thermal, chemical, and mechanical properties. Traditionally, glass was used for the manufacture of decorative items and craft, but with easy workability, coupled with exceptional transparency, strength, transmittance, and U-value (i.e., heat conductivity), glass is increasingly used in high-end furniture.

The ultimate strength of glass is related to the rate at which it is cooled, and this differentiates the material into four types, i.e., (1) **annealed glass** is the lowest grade of glass, produced when the molten glass is allowed to cool in a controlled manner to relieve stresses, (2) heat-strengthened glass is the result of reheating the **annealed glass** up to about 700 °C and then cooling it quickly, to toughened it further, (3) **tempered or toughened glass** is the most common type of glass used in table tops and other load-bearing furniture. It is produced by heating annealed glass to 700 °C by conduction, convection, and radiation, which creates different cooling rates between the surface and the inside of the glass, resulting is highly toughened glass, and (4) laminated glass is produced by laminating two sheets of toughened glass with polyvinyl butyral (PVB) interlayer. Laminated glass offers many safety and security advantages as high shattering impact. In this context, it is no surprise that glass in increasingly becoming a material of importance for furniture manufacturing, especially for exquisite furniture. For further details on glass as a material for furniture, the readers are referred to Musgraves et al. (2019).

Stage 3: The rough milling section performs two main functions: (1) it eliminates defects from the raw material and (2) it produces blanks or squares of specific length, width, and thickness for further processing (Fig. 4.4).

In essence, the rough milling has a significant implication on the yield/recovery of the raw materials used in the production of furniture, which in turn affects the product manufacturing costs.

Fig. 4.4 Production of blank or square

Fig. 4.5 Production of profiled component

Three important machining operations are usually characteristics of the rough mill, namely cross-cutting (cut-to-length), thicknessing (cut-to-thickness) and ripping (cut-to-width). A combination of the thicknessing and ripping operations into a single operation is known as 4-siding, which leads to surfaced-4-sides (S4S) components. This operation is usually carried out using a 4-sided molding machine.

Stage 4: The next stage is the machining shop, which produces components of specific shapes/profiles using the blanks from the previous section (Fig. 4.5). A mark-out of the shape/profile is done initially on the blank, followed by the rough cut-out using the narrow band saw. This rough cut-out is then machined with greater accuracy using a spindle shaper and finished off with a high-speed router. The final white piece of the component has the correct dimension as well as shape/profile as per the product drawing or prototype. The only allowances or tolerances given to these components are for the subsequent jointing and sanding operations.

In more advance manufacturing environments, the CNC work centers are placed in this section to perform multiple functions quickly and accurately on the components.

Stage 5: The individual whitewood components are then machined to incorporate its joint configuration (Fig. 4.6). The common types of furniture joints are (1) dowel joint, (2) mortise-tenon joint, (3) dove-tail joint, (4) finger joint, and (5) miter joint. The type of joints used in a piece of furniture is dependent on the structural strength required, joint design, construction technique as well as the material used. The

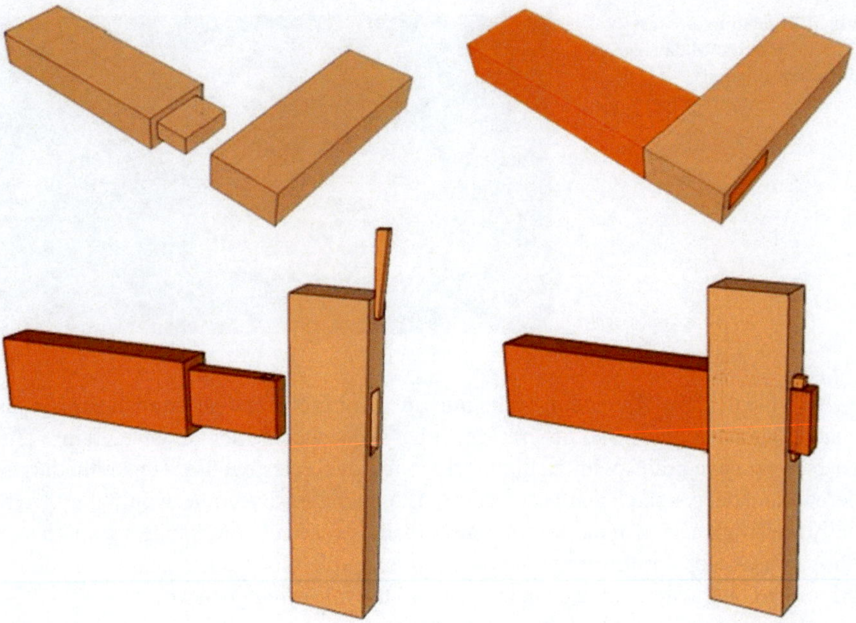

Fig. 4.6 Joints in furniture

machining of these joints is carried out using the boring machine, mortiser, and tenoner, shaper including finger shaper or router, respectively.

The machining of the joints must be accurate and precise to avoid the formation of loose joints, eventually leading to a weak structure. These joints are usually glued together using common furniture adhesives, and the glue-line must necessarily be of minimal thickness after curing or hardening. A thin glue-line is essential to transfer the load from one member to another, with minimal contribution to the overall strength of the joint. It is for this reason that joints must be machined accurately and precisely to avoid a thick glue-line that leads to lower joint strength.

Stage 6: The abrasive sanding operation smoothens and flattens, or evens-out the surfaces of shaped/profiled components, or even the sub-assemblies. In essence, sanding is the pre-finish operation which lays the foundation for the application of finishes or coating (Fig. 4.7).

The term sanding refers to the use of the sandpaper as the abrasive medium. However, there are several types of abrasives in the market. During the sanding operation, sharp abrasive particles or grains are forced into the surface of the workpiece, which leads to the removal of small bits of the workpiece. It is comparable to a scarping-action, where each abrasive particle act like a miniature cutting tool. Since the abrasive grains are much smaller than the conventional cutting tool, and its' geometry and orientation not well defined, abrasive sanding as a process is inefficient in power consumption, which is mostly generated as heat (Ratnasingam et al. 2004).

Fig. 4.7 Sanding process (Courtesy of the Malaysian Furniture Council)

A typical abrasive sanding schedule in furniture manufacturing begins with coarser abrasive grains and progresses with a sequence of finer abrasive grains. The coarser abrasive grains with its higher stock removal rates, would remove the larger defects and deep scratch marks and dents. The finer abrasive grains would then gradually remove the scratch marks previously made, eventually producing a smooth and flat surface ready for further processing. An example of a typical sanding schedule prior to wood finishing may be in the order of the following grit-sizes: 80–100–120–150–180. The sanding operation is carried out using a variety of machines, depending on the specific surface types and profiles. Among the common sanding machines used include the wide-belt sander, edge-sander, stroke-sander, brush sander, and the hand-held orbital sander.

Stage 7: The finishing operation serves three important purposes: (1) it improves the esthetic appeal of the furniture, (2) it protects the surface of the furniture, and (3) it allows equalizing or camouflaging the original color of the furniture to suit the customer's desire. The finishing operation in furniture manufacturing must be carefully planned and executed to ensure that the furniture looks appealing, free of defects, meets the customer's expectations, while conforming to safety and environmental requirements (Fig. 4.8).

Furniture finishing begins with a thorough inspection of the pre-sanded components or sub-assemblies. Imperfections, holes, dents, and pin holes are usually filled up and concealed using putty or fillers. It is then scrapped and sanded by hand using a sanding block or a hand-held sander. This is followed by the toning process, if necessary. The surface color is altered or changes either by bleaching or staining (i.e., using dye-stain or pigment-stain, depending on the clarity required). The surface is then applied with several coats of finish, with sanding in between the layers to improve adhesion. Many different types of finishes are available in the market, including *shellac*, varnish, *drying oil*, *lacquer*, or *paint*, and the choice used is dependent on the desired quality and customer preference. The finish is usually allowed to dry after each consecutive layer to achieve the necessary inter-coat adhesion and finish film strength, and each coat is typically followed by sanding. The final finish may be *polished* or buffed using *steel wool*, *pumice*, or other materials, to give the surface

Fig. 4.8 Finishing operation

a degree of shine. Sometimes, the customer will require a mild coat of wax to be applied on to the finish surface for added protection.

Although there are many methods of applying finishes to furniture, such as dipping, brushing, roller coating, and spraying, the latter is still the most widely used method. Spraying is a technique in which a spray gun applies the *coating* material (varnish, lacquer, paint, etc.) through the principles of atomization onto the surface of the furniture. Compressed air is used to atomize and direct the coating particles onto the surface of the furniture. Such application method of coating materials results in an even yet uniform coating film on the furniture. With developments in spraying technology, this technique has also become increasingly efficient. The coating applied on to the furniture is left to harden or cure, in ambient condition, or in a drying oven for accelerated curing before the furniture is sent for packaging.

Stage 8: Upholstery in furniture is an important value-adding activity, but it is not applied to all furniture. It refers to attaching furniture frames with cushioning, padding, and textured material covers, which not only increases the appeal of furniture but also provides superior comfort. It is a skilled job, and modern upholstery uses synthetic foam, spring, and fabric, rather that wool, animal hairs, etc.

Stage 9: Packaging serves a multitude of purposes, namely: (1) it provides physical protection against mechanical damage, (2) it provides barrier protection against the elements, such as moisture, oxygen, etc., (3) it helps out with containment or agglomeration of many components that goes into a piece of furniture, (4) it assist with information transmission pertaining to the assembly of furniture, carbon footprint, recyclability, etc., (5) it help with marketing of furniture, (6) it helps with the brand positioning, (7) it improves the security during transportation and distribution, and (8) it improves convenience during handling and transportation. It must be recognized that the packaging design is a science.

A variety of packaging materials are available for furniture, but the primary considerations are often ease of operation and cost. The most common are corrugated boards and cartons on the exterior, with polyethylene (PE) foams, bubble wrap, and machine and hand stretch films as interior padding (Fig. 4.9). The exterior packaging is then secured using plastic strapping and adhesive tapes. Proper packaging of furniture is important to ensure that the furniture reaches the customer in pristine condition.

Against this background, it is apparent that furniture manufacturing is indeed an art as well as a science, where intricate processes are implemented in stages until the final piece of furniture is made, and ready to be shipped to the customer. Along the manufacturing processes, opportunities for errors and mistakes are aplenty, and therefore close control of the operations based on scientific and engineering principles will ensure a high productivity manufacturing operation (Scott 2008; Huang and Intarakumnerd 2019). It is for this reason that the science and engineering facets of

Fig. 4.9 Furniture packaging

furniture manufacturing must be appreciated to enable the furniture maker to take charge of all operations, and not leave it to chance.

Summary

- Furniture manufacturing begins with product ideation and prototype making.
- It then goes through a series of value-adding processes, such as machining, jointing, sanding and finishing, until the final piece of furniture is made.
- It is then packed and sent to the customer.

References

Black JT, Kohser RA (2019) DeGarmo's materials and processes in manufacturing, 13th edn. Wiley, Hoboken, New Jersey, USA

Cybulski E (2009) Plastic conversion processes—concise and applied guide. CRC Press, Boca Raton, Florida, USA

Desch HE, Dinwoodie JM (1996) Timber—structure, properties, conversion and use, 7th edn. Macmillan Press Ltd., Basingstoke, Hampshire, United Kingdom

Huang YL, Intarakumnerd P (2019) Alternative technological learning paths of Taiwanese firms. Asian J Technol Innov 27(3):301–314

Musgraves JD, Hu J, Calvez L (2019) Handbook of glass. Springer International Publishing, Cham, Switzerland

Ratnasingam J, Scholz F, Friedl E (2004) Wood sanding processes—an optimization perspective. UPM Press, Serdang, Malaysia

Ratnasingam J (2013) Optimization of furniture manufacturing processes. Tech. Note No. 17, IFRG Publication, Singapore

Scott AJ (2008) Patterns of development in the furniture industry of Thailand: organization, location, and trade. Reg Stud 42(1):17–30

Stokke DD, Wu Q, Han G (2014) Introduction to wood and natural fiber composites. Wiley, Chichester, West Sussex, United Kingdom

Zehui J, Zhenhua P (2007) Bamboo and rattan in the world. Chinese Forestry Publishing House, Beijing, China

Chapter 5
Preparing the Furniture Parts

The machining of furniture parts and components serve two main purposes, which are to machine the final profile and shapes required of each component in the furniture, while at the same time ensuring that the final dimensions of the components are as per the design specification. Unlike the rough milling operations, in this section, the machining is more intricate involving specialized tools. Machining precision and accuracy are important process requirements. The cutting speeds and the amount of fine dust generated is much higher in these operations. This chapter describes the main processes and machines involved in the preparation of furniture components based on solid wood as well as wood-based panels.

5.1 Introduction

In furniture manufacturing, the machining of wood is undertaken through a mechanical process, where the cutting tool removes chips from the wood or workpiece. The machining or cutting principles are almost similar for every operation in the woodworking industry, but the cutting tools used are markedly different depending on the specific operation or machine used. During the machining process, the cutting knives attached to the rotating cutterhead that separates the workpiece material as individual chips. The feed rate is limited by the size and type of cutting tool used. The cutting process is intermittent and the cutting knife alternatively cut short chips of varying thickness. Generally, wood and wood-based material are machined in all directions, but most often along the direction of the wood fibers. The direction of cut is usually chosen to be conventional, i.e., against the direction of feed. In theory, two categories of machining processes are encountered based on the axis of rotation and the surface created by the cutting edges. If the process uses a cylindrical tool axis which is parallel with the workpiece surface, it is known as orthogonal cutting. On the other hand, if the front axis of the tool is perpendicular to the workpiece surface, it is known as peripheral machining (Koch 1964).

J. Ratnasingam, *Furniture Manufacturing*, Design Science and Innovation,
https://doi.org/10.1007/978-981-16-9412-7_5

The productivity of the machining process is usually a function of the tool life. Tool life is the period during which the knife cuts effectively, and it simply refers to the time for the tool edge to lose it sharpness and become dull. Dulling of tool, also known as tool wear is influenced by tool geometry, tool material, and the machining process parameters. When a dull or worn-out tool is used, the cutting process becomes less inefficient, consumes more energy, increases frictional force, and the resultant surface quality deteriorates. The reverse is true when the tool is sharp (Ratnasingam and Tanaka 2002). The mechanics of the wood cutting process is the interplay between workpiece characteristics, tool material and its geometry, and the cutting parameters used. Therefore, in wood machining processes, optimization is highly desirable as it finds a compromise between production cost and quality (Ratnasingam et al. 1999a).

Cutting tool materials must be harder than the wood material which is to be cut, and the tool must be able to withstand the heat and force generated in the cutting process. Therefore, cutting tool materials must be hard, tough, and with good wear resistance. The common cutting tool materials used in the industry are high-speed steel, cemented carbides, ceramic, and polycrystalline diamond, although high speed steel and cemented carbides predominate. Further, the tool must have a specific geometry, characterized by the tool's rake/hook, sharpness, and clearance angles. The clearance angle in the tool is usually designed so that the cutting edge can contact the workpiece without the rest of the tool dragging on the workpiece surface. The tool geometry may also incorporate lateral relief and shear angles to facilitate the cutting process. In essence, to prolong the tool's working life, all of the above factors must be optimized, especially the process parameters, such as the chip thickness, and speeds and feeds at which the tool is run. It must be emphasized that when the stock removal rate increases, the work done during the machining process increases proportionately, influencing both the surface quality and production cost (Ratnasingam and Tanaka 2002).

Cutting tools can also be differentiated as tools with linear, rotary, or a combination of both motions. The band-saw and lathe are tools with linear motions, while the drill bits, router bits, and planer heads are examples of tools with rotary motions. Cutting tools are also classified as single-point cutters, or multi-point cutters. An example of the former is the turning lathe, while the router bits and planer heads are considered multi-point cutters. Cutting tools may also be categorized as either solid-cutters, i.e., the cutting edge is part of the tool body, or tipped-cutters, where the cutting edge another material which is brazed, welded or clamped on to the tool body. In this respect, wood machining processes is rather challenging as wood and wood-based materials are natural variable material, and therefore in-depth deliberation of the subject is beyond the scope of this book. Nevertheless, for a detailed discussion of the fundamentals of wood machining processes, the readers are referred to the works by Koch (1964), Ratnasingam and Tanaka (2002), and Csanády and Magoss (2013). Figure 5.1 shows the wide array of cutting tools used in the woodworking operations.

The different machining processes in furniture manufacturing share a common chip formation process, and this relationship that defines the cutting process, has enabled a better understanding of the different wood machining processes (Koch, 1964). The notable difference is the abrasive sanding processes, which has a negative

Fig. 5.1 Different types of woodworking tools (Courtesy of the Malaysian Furniture Council)

rake angle, and the cutting edges are randomly aligned with specific patterns. The 'scraping' action typical of the abrasive sanding process, is therefore inefficient in terms of energy consumption, as much of it is released as heat due to friction (Ratnasingam et al. 1999b).

Woodworking machines are designed for heavy-duty applications, as the cutting processes involves significant amount of energy. The work done during the removal of a single chip during mechanical wood machining processes is fundamental (Ratnasingam et al. 1999a). This process is energy consuming since the cutting knife needs to overcome the resistance of the work piece and the inherent friction involved. However, the energy consumed in a single chip removal during mechanical wood machining usually involves a fraction of a kilowatt (kW) of energy, but since the cutters are rotating at high speeds, it requires significantly large amounts of energy (Koch 1964). To ensure efficiency, most woodworking machines are designed with three important elements, i.e., an electric motor, that supplies the energy, which is transmitted by a drive pulley to the socket that holds the cutter, which is then rotated to perform the machining operation. For further discussion on woodworking machine design, the readers are referred to Collins et al. (2010) and Jiang (2019).

In executing the cut, most wood cutting tools use a free-cutting mechanism, and to produce a clean cut, high cutting speed and sharp edges are necessary (Csanády and Magoss 2013). Nevertheless, with their comparatively lower strength compared to hardwoods, softwoods can be worked at lower cuttings speeds without sacrificing surface quality. In furniture manufacturing, the types of machines used is predetermined by the nature and complexity of the parts or components to be machined. It is notable that in many of the SMEs in the furniture industry, simple, stand-alone woodworking machines performing only one operation (sawing, planing, shaping, jointing, abrasive sanding, etc.) is common. The use of multi-head planers accelerates the machining of parts and components, since two or more surfaces can be machined in a single pass, thereby reducing machining time significantly (Ratnasingam and Tanaka 2002).

For high-capacity furniture factories, manufacturing lines with different woodworking machines and even special machining units are laid-out. An example of this is in the manufacture of edge-banded panels for bed or table production. Such 'process layout' are typical for high-volume production factories that are engaged in the mass-production system (Merdzhanov 2018). When smaller production volumes are required, most often with product diversity, the machines are arranged in a 'product layout', whereby machined with similar functions are grouped together in one section of the factory shopfloor, or the use of CNC machines equipped with different cutting heads becomes the obvious manufacturing solution, to accommodate the different sizes and profiles of components (Csanády et al. 2019).

In furniture manufacturing operations, the key to a productive machining operation is to ensure that the most suitable cutting tools are selected for a particular machining task, depending on the nature of the workpiece to be machined, and the resultant quality. Off course the tool, machining, workpiece characteristics all interact to result in the final machining outcome, it is crucial for the furniture manufacturer to take these factors into consideration prior to the start of the manufacturing processes (Ratnasingam and Tanaka 2002). The subject of tool materials and its wear characteristics together with the interaction of machining process parameters have been extensively research, and a detailed discussion provided in Csanády and Magoss (2013). The various mathematical relationship involved in the machining processes is the subject of Chap. 14 in this book. Nevertheless, it must be emphasized that a compromise must be reached between the highest production rate and the lowest production cost, to ensure that the furniture manufacturer is able to meet his production targets without suffering too much in terms of quality and cost. Under some circumstances, absorbing a higher production cost, could result in significantly higher production rates, which make it rewarding for the furniture manufacturer (Csanády et al. 2015).

5.2 Rough Milling

Furniture manufacturing usually starts with the rough milling. The dried sawn timber or sawn lumber (usually with a moisture content below 10%), is sent to the rough mill, where it is first sorted out by dimensions and grades. This sorting is done to facilitate the machining operation, while ensuring that the highest yield or recovery is obtained. Rough milling is the first machining operation within the furniture manufacturing factory. The rough sawn lumber, usually stored in the store or lumberyard, is put through the rough milling operation. The sawn lumber is transformed into blanks/squares parts, with length, width, and thickness necessary to produce the required furniture part, which may be a chair leg or other components (Csanády et al. 2019). Since rough sawn lumber are usually over-sized or with an allowance, the primary objective of the rough mill is to produce blanks/squares with the correct dimensions to process it further into components in the next section. The secondary objective is to ensure that the yield or recovery from the sawn lumber is maximized by eliminating defects, including inherent wood defects, and drying defects, from it (see Appendix for a list of common defects). Therefore, rough milling in essence has the highest cost implication in furniture manufacturing, as it affects the wood raw material recovery in the factory (Ratnasingam and Tanaka 2002). Before going further into the rough milling operation, several terms must be clearly defined.

- *Sawn timber*—Raw material procured from the sawmill in the form of beams and planks.
- *Blank*—A rectangular or square shaped part or component, usually with four machined surfaces, expect the ends, that goes into the machine shop.
- *Rough dimension*—The measurement or size of the blank/square.
- *Finished dimension*—The measurement or size of the machined furniture part or component.

Rough and Fine Machining.

The difference between rough and fine machining is the amount of stock removed. In the case of rough machining, up to 2.5 mm can be removed, while in fine machining, small stock removal of about 0.5 mm–1.0 mm takes place. Generally, machining a piece of wood is divided into two levels of roughness (Csanády et al. 2019), the first level is aimed at removing large chunks of wood material to even out the rough surfaces of the rough sawn lumber. In many instances, natural wood defects, such as knots, tear-out, chip mark, etc., and drying defects, such as split, honeycomb, etc., may be exposed during this rough machining as large ships are removed (Kilic et al. 2006). Therefore, the incident cutting forces in this operation is equally large (Koch 1964).

This is followed with another rough machining pass, which evens out the surfaces further. The fine machining is aimed at producing the components with the desired surface quality, with minimal defects (Stewart 1970).

These two machining operations requires different cutting tool parameters. Rough machining is usually carried out with cutters with negative rake angles, which allows higher cutting speeds and higher cutting forces. On the other hand, fine machining often uses cutters with positive rake angles, which produces better surface quality, but limits the amount of stock that could be removed (Ratnasingam and Tanaka 2002).

The sawmills usually cut sawn lumber of various lengths, and cross-sectional sizes, as defined by the market requirements, as well as to improve yield. Inevitably, sawn lumber comes to the rough mill in furniture factories, in random dimensions, or sizes. The sawn lumber is initially graded and sorted out by its quality grades and dimensions and bundled up according to the wood species. At the rough mill, the actual cutting plan is worked out to produce the quantity of blanks/squares required for a production order. The finished dimensions of the various furniture parts required for a piece of furniture is shown in the bill of materials (BOM) and cutting plan for the furniture. Once the finished dimensions are known, it is easier the work out the best rough size sawn lumber required, based on widths and thicknesses of the available stock in the lumber yard. In many countries throughout the world, there are common cutting sizes produced by the sawmills, and therefore furniture manufacturing must adopt to these available sizes in the market, to ensure economical manufacturing process (Csanády et al. 2019). Therefore, the primary function of rough milling in the furniture factory is to select and use the most suitable sawn lumber, in terms of its sizes and grades, (i.e., *also known as the common cutting sizes in the sawmills*), for a given product, and to cut the sawn lumber into specific dimensions for the further processing into the various furniture parts and components (Fig. 5.2).

As mentioned previously, the secondary role of rough milling is to produce furniture parts, or components that are free of common wood defects and checks and/or splits. The rough sawn lumber may have natural wood defects, such as knots, wane, rot, and worm holes, and the presence of these defects will impair the quality of furniture manufactured. Therefore, these defects must be identified, and eliminated

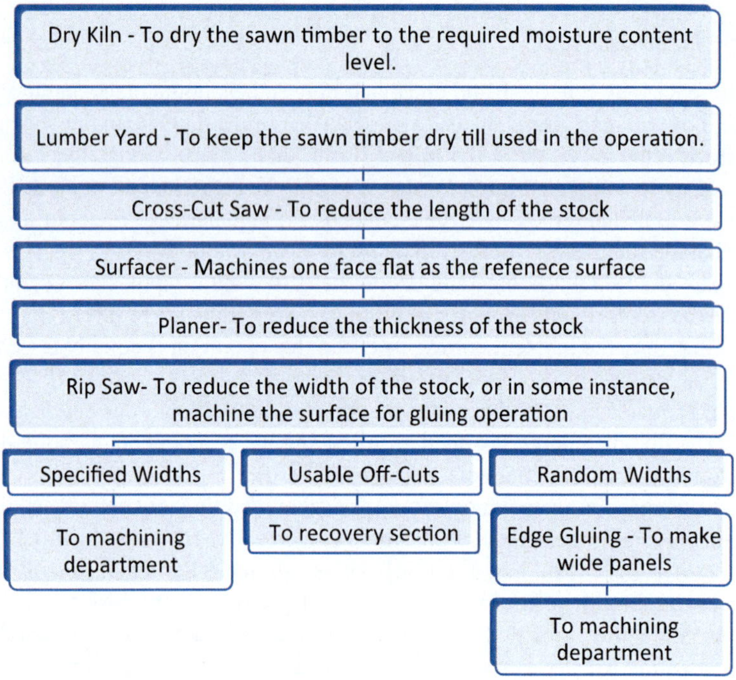

Fig. 5.2 Rough mill operations (Adapted from Csanády et al. 2019)

in the rough milling section, and the blanks/squares moving into the next section are free of such defects and have clear surfaces (Ratnasingam et al. 1999a). However, a few minor defects are acceptable in concealed, or hidden parts of the furniture, which have minimal impact on the overall quality and product lifespan. A good example of such defects is pinholes found in some timbers. In this regard, the rough mill is responsible to convert the sawn lumber into blanks free of defects of the desired dimension. These two functions must be combined to maximize the output of blanks/squares, with the least amount of input rough sawn lumber (Csanády et al., 2019).

According to the study by Ratnasingam et al. (1999a), the inefficient rough mill can add up more than 4% to the raw materials cost, especially if the improper sizes of rough sawn lumber are used in the rough milling. Another 8% cost may be added, if defects are not eliminated in the rough mill, parts, and components with such defects are processed further. An important economic point in view of both these objectives would be to conduct these operations with minimum material waste and labor cost.

Another problem that is commonly encountered in the rough mill is the processing of warped sawn lumber. As a result of uneven shrinkage, the boards are 'curved' or 'warped', and a surfacer is the machine used to produce a flat surface. The function of the surfacer is to machine the rectangular piece of stock, with one level face, as reference for future machining operations.

The typical machines found in the rough milling section in the furniture factory are as shown in Fig. 5.3. The rough sawn lumber is initially cut to length using a *cross-cut saw*, which determines the optimal length required for the furniture part or component. This is followed by the *thicknesser* or double surface planer, which machines the top and the bottom surfaces simultaneously to the specific thickness. In some instances, a single surface planer is used for this purpose, which also produces the necessary reference plane for further machining. Finally, the width of the sawn lumber is machined to the desired width using a ripsaw. Surfacer/planer can machine one surface or two surfaces, simultaneously, but there are multi-head planer machines, with four or more cutting edges. Such machines, known as *molders*, or *4-sider*, and are built in a way they can handle stock of different widths up to the opening width of the infeed table. The workpieces are fed into the machine, one piece at a time, or a few pieces edge to edge. Molders are considered the 'workhorse' in the furniture manufacturing factory due to their high outputs.

A well machine-planed lumber allows the operator to pick up small defects and blemishes easily, and with its unform thickness, it is in a better form for edge gluing to produce larger panels (Davim 2013).

It should be emphasized that the molding machine is a through-feed machine, with a few consequences. The moving workpiece vibrates, and the amplitude of the vibration have a strong impact on the resultant surface finish, smoothness, and dimensional accuracy. Infeed and outfeed rollers are prominent features of this machine to minimize these effects. One inherent feature of molding operation which is obvious to the naked eye, are the 'cutter-mark' or the wavy surface caused by the cycloid direction of travel of the knife. If the waviness on the surface is obvious, reducing the feed

Types of Machine	Function
	**Cross Cut Saw/Jump Saw** A crosscut saw is a saw that cuts the sawn lumber across its grain. It is the process of cutting-to-length.
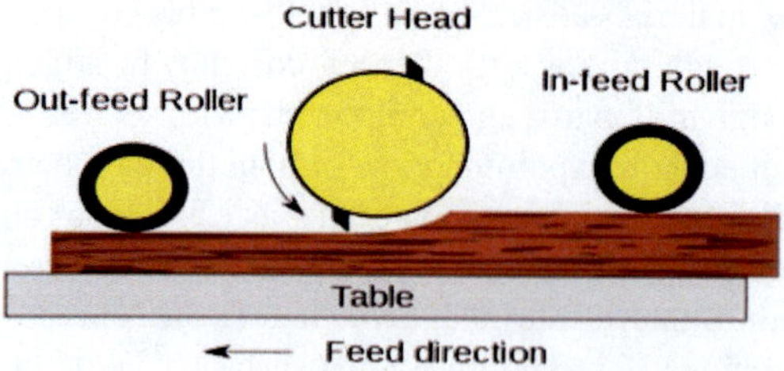	**Thicknesser/Planer/Surfacer** It is cutting- to-thickness based on a reference plane made initially by the planer or surfacer. This operation is also known as 'deep cutting'.
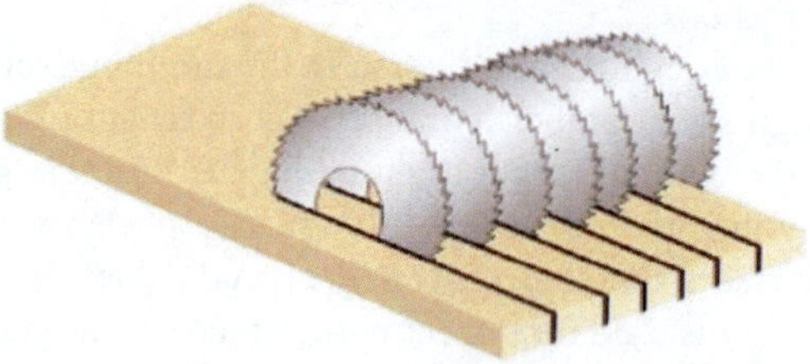	**Rip Saw** It is a saw that cuts along the grain of the sawn lumber and it is cutting-to-width. Care should be taken to avoid kick-back of the workpiece by attaching a riving knife. This operation is also known as 'flat cutting'.
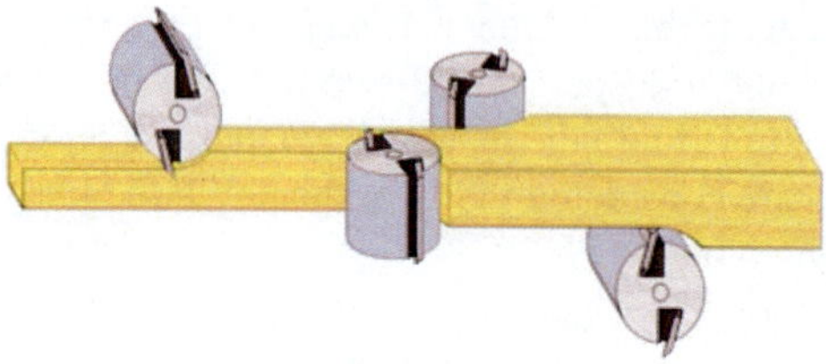	**Molder/4-Sider** It can machine all for surfaces simultaneously to produce surfaced-4-sides (S4S) components, or the cutting heads can be fitted with specific profiled cutters to produce finely machined profiles in one pass. The latest molders have up to 8 or more, cutting heads and can produce micro - finish of high-quality surface.

Fig. 5.3 Common machines in the rough mill (Courtesy of the Malaysian Furniture Council)

speed minimizes its effect, but at the expense of the throughput rate (Ratnasingam and Tanaka 2002).

At the end of the rough mill, the blanks/squares of specific dimensions, and free of defects, for the furniture parts, is ready in the desired quantity to be moved into the next section.

5.3 Machine Shop

The blanks/squares from the simple squaring operations of the rough mill, are brought into the machine shop, which is the next section in furniture manufacturing. Here, the blanks/squares are converted into components/parts of various profiles and forms, which resembles the final form of the furniture. This conversion is achieved through a series of machining processes, commonly known as 'finish machining' (Csanády et al., 2019). Unlike the rough milling section which focuses on producing squares with straight, flat, and defect-defect-free surfaces, the machine shop is focused on imparting a geometric alteration to the square, into shapes and profiles for the furniture.

In machining furniture components/parts, the order of the machining process is usually pre-set in the production planning department. This is often based on available machine capacity and technical specifications dependent on the final profiles and shapes to be machined. Nevertheless, there may arise moments when the prescribed machining sequence cannot be followed, such as during machine downtime or unforeseen maintenance, which requires a change in the machining sequence. However, in most instances, it is best to follow the prescribed machining sequence to optimize production, which ensure maximum throughput rate, acceptable quality at minimal cost (Ratnasingam and Tanaka 2002). Usually there is one specific order in which the machining is carried out. There may be instances to switch the arranged sequence of operations, but for the most part sticking to the pre-arranged machining, sequence is way better to ensure the throughput rate and quality are achieved.

Among the factors considered in determining the machining, sequence are (1) the availability of a reference plane, (2) intricate profiles and shapes, (3) desired quality, and (4) throughput rate (Csanády et al. 2019). The availability of the reference plane on the blanks/squares, allows the pieces to be processed in a machine that requires close machining tolerances and guiding surface. Every machining operation for the furniture components/parts must be taken in reference, to one or two planes, which will ensure the precision and accurate machining processes. The intricate profiles and shapes mean the components/parts must pass through the specific machines to achieve the necessary profiles. The desired quality, both in terms of defect-free surfaces and smoothness are important in profiling and shaping work, that any sacrifice in this quality, will be detrimental to the overall quality of the furniture (Ratnasingam and Scholz 2007, 2008). In this context, the sequence of machining will ensure that the profiles and shapes are incrementally added to the blanks/squares, while maintaining quality. Finally, the machining sequence is often determined through work study or industrial engineering measurements, which establishes the available capacity in each machine station, and works out the shortest route through the shopfloor for the component/part. In essence, the machining sequence for furniture components/parts through the shopfloor is a matter of production economics (Ratnasingam and Tanaka 2002).

As mentioned previously, 'work study' or 'industrial engineering' is carried out in furniture manufacturing to establish the data on machining capacity, machine setup time, anticipated tool change time, possible machine downtime, and actual machining time (Muhammadi et al. 2020). This information is important for the development of the Master Production Schedule (MPS) which provides an overview of the production schedule for a given period-of-time through the factory. It serves as a timetable of activities throughout the factory, while tracking the movement of components/parts, known as work-in-progress (WIP) through the factory shopfloor. Based on the MPS, daily route sheets, and operational charts are worked out to allow close monitoring of progress on the factory shopfloor.

Several operations are commonly experienced within the machine shop, which incorporates rough cut-out, shaping/forming, routing, and turning. In some factories, this section also houses the CNC machines and workstations, which carries out various machining operations (Fig. 5.4).

- In the machine shop, the first operation is often marking out the shape of the component on the bank/square. This serves as the reference for the next operations, which is the rough cut-out.
- Rough cut-out is carried out using the narrow band saw, where the width of the band saw is often 25 mm or less. The primary purpose of this operation is to cut out the shape based on the marking out previously done. Two types of operations can be done on the narrow band saw: (1) straight-cut, freehand-cut, or cutting with the fence as the reference, and (2) curved-cutting, which is often carried out with the bank or square mounted on to a jig/template, for accuracy and repetitive cuts. In all instances, the rough cut has a large allowance of up to 3 mm, which must be machined away from the profile in subsequent operations. Since the rough cut-out is an important process, several parameters must be considered to optimize the operation. The width of the saw blade, blade tensioning regime, blade stability and vibration, cutting allowances, energy consumption, and its cutting accuracy are factors that must be borne in mind (Csanády et al. 2019).
- Shaping, as the name implies, is the process of shaving off the extra allowance from the rough cut-out components/parts from the previous machining operation. It also imparts the actual profile and shape desired in the component/part. The shaping operation is often carried out using a single-spindle shaper, double-spindle shaper, or an automatic multi-head shaper. A shaper is used for larger cuts, removing up to 3 mm from the rough cut-out components/parts, which makes the actual profile more visible with clean cuts. The heavy-duty nature of the shaper makes it a machine with a large capacity, and if used with jigs/templates it can accommodate large volumes of components/parts very efficiently.

Types of Machines	Function
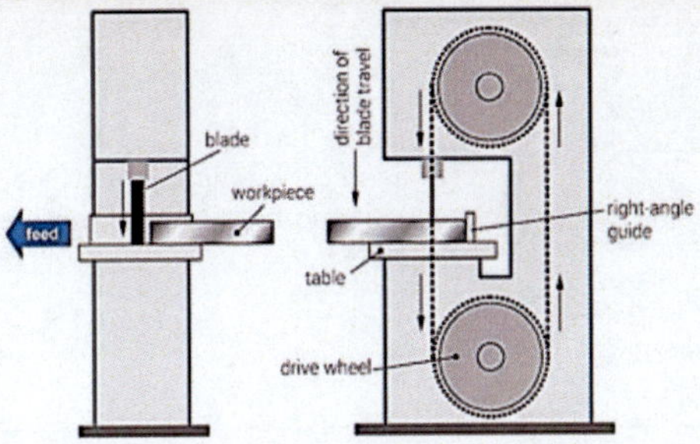	**Narrow Band Saw** It is a versatile machine that is widely used in furniture manufacturing. It is used most effectively to make rough cut-outs of the shape/profile.
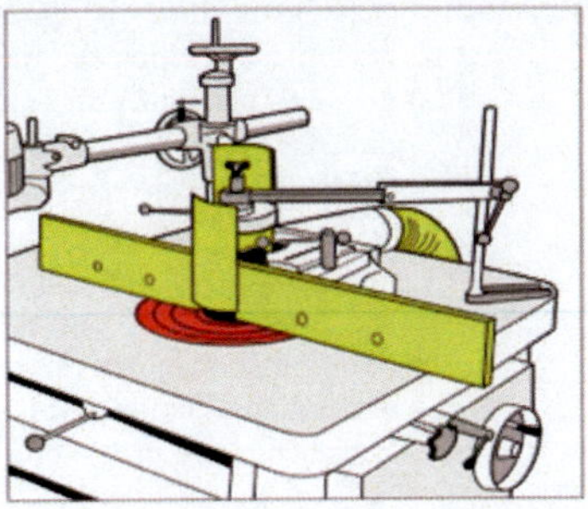 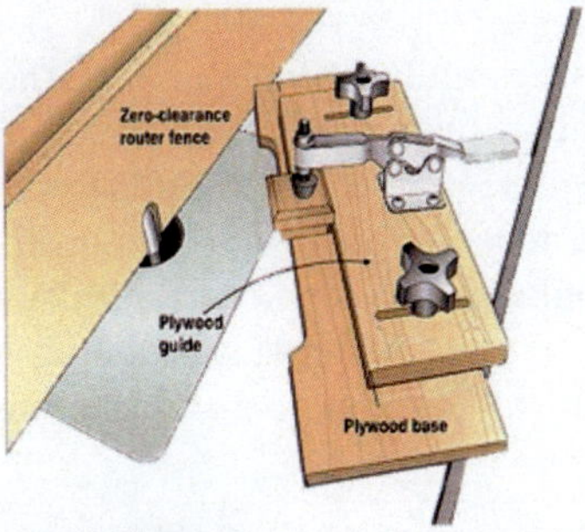	**Single or Multiple Spindle Shaper** One of the most versatile machines in woodworking, it can be used for molding, rebates, and curved work.
	**High Speed Router** Used for rebates and moldings, generally one off-jobs rather than using a spindle molder. Material is fed towards the direction of the cutters, and the four cutting edges produce finer surface finish than the shaper because of its higher cutting speeds.

Fig. 5.4 Common machines in the machine shop (Courtesy of the Malaysian Furniture Council)

<table>
<tr>
<td>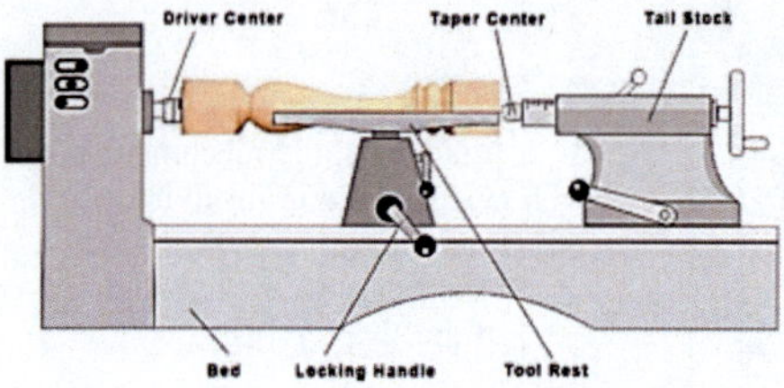
</td>
<td>Turning Lathe
It is used for turning out special circular profiles/shapes from squares.</td>
</tr>
<tr>
<td>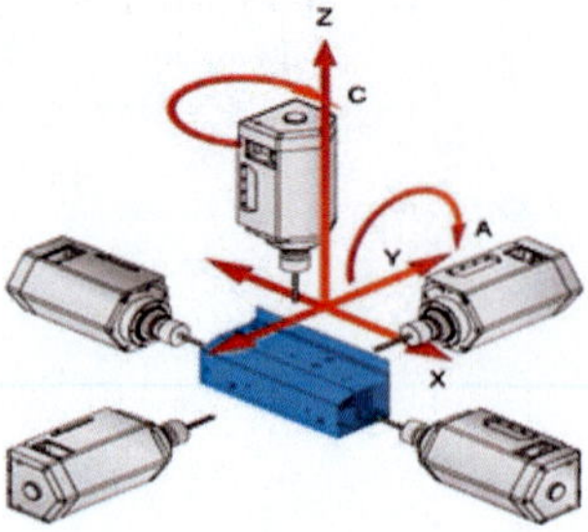
</td>
<td>CNC Machine/Workstation
The computer numerical control (CNC) machine is an advanced machine, in which the movements of the machine and tools are dictated by a pre-programed computer. It can carry out several different machining operations, including sawing, shaping, routing, and boring. CNC machines can accomplish complex three-dimensional cutting tasks as the tools can move in 3, 4 or 5 axes.</td>
</tr>
</table>

Fig. 5.4 (continued)

- The final process in the machine shop is the routing operation. This is the finishing up of the profiles and shapes in the furniture components/parts. With small diameter profiled cutters at high speeds, the router is able to rout/hollow out, chamfer, or impart intricate shapes on to the components/parts with high quality. Router cutters can have intricate, complex profiles, which can impart distinctive shapes, profiles, and smooth cuts that makes the furniture design stand out. Both hand-held and machine-aided routers are common in the furniture manufacturing industry (Csanády et al. 2019).

The application of computer-numerical controlled (CNC) machines and workstations are increasing throughout the furniture manufacturing industry. Operating at high speeds, with multiple cutting heads to perform various machining operations, it offers flexibility, in both production volume as well as product diversity (Koc et al. 2017; Bagnall 2021). Initially launched as machines working in 3-axes, the latest machines are capable in working in 4, 5, and 6 axes, making the manufacturing of complicated profiles and shapes rather easy. The coding of the machine program into the computer, which controls both the machine and tools movements, enables the accomplishment of such tasks, with accuracy and precision. Once the cutting pattern is programed into the computer, it is a machine that can work round the clock, and with automatic loading and unloading attachments, the throughput rate from such machines is very high. In fact, investing in a CNC machine can replace the need for many other stand-alone machines, which not only saves space on the shopfloor but also improves overall productivity. Nevertheless, the decision to either invest in a CNC machine, or customized/specialized through-feed machine is predetermined by the production volume desired and the diversity of products to be manufactured. Generally, in flexible manufacturing environments, CNC machines are used to produce large volumes of components/parts (Csanády et al. 2019).

In some furniture factories, lathes are used to produce turnings of specific profiles, for bedroom, chair, or table components/parts. Wood lathe are machines, that turns a profile out from the blank/square, using a chisel guided by template, which serves as the reference profile. Previously manually operated, the newer lathe machines are automatically operated, increasing their throughput rate significantly (Conover 2019). Due to the high stresses applied, the blanks/squares used for turnings must be free of internal defects, to avoid breakage during turning.

Machining accuracy in the woodworking industry is determined by many variables, including vibration character, rigidity, excitation force, and the workpiece clamping system of the machine. Furthermore, the cutting parameters, the accuracy of the cutter head running circle, and knife sharpness will not only impact the resultant surface quality, and dimensional accuracy, but also the effective life of the knife/tool. Poor surface quality and dimensional accuracy are usually associated with a worn-out tool /knife edge, which produces higher amplitude vibrations, during the successive knife engagements with the workpiece. The deviation in machining causes the need for strict allowances and is the basis for Statistical Process Control (SPC) applications in the furniture manufacturing (Simanová and Gejdoš 2015).

The dimensional accuracy of a machined work piece can be characterized by the variation between the nominal measurement (design-value) and observed measurement due to machining, and by the resultant standard deviation (σ) between these two measurements. Previous research has shown that in wood machining processes, the allowable tolerance which corresponds to 3σ, which is the limit of the standard deviation, is usually acceptable. Therefore, under normal circumstances, with a probability level of 0.275%, only a quarter per cent of work pieces in a batch, may have higher tolerances than its prescribed value (Csanády et al. 2019).

In this context, efforts must be taken to ensure minimal vibrations occur when the components/parts are being machined. Pneumatic and vacuum-suction clamping systems are widely used in the woodworking industry, to provide reliable clamping, especially on CNC machines. Such clamping system have proven to be more reliable than through-feed machines, as a moving workpiece tend to vibrate especially when it leaves the engaging cutter head towards the rollers. Under such circumstances, chatter-marks, roller marks, and cutter marks become apparent, hence sacrificing quality as well as the machining accuracy (Aguilera et al. 2000).

5.4 Differences Between Rough and Fine Machining

Machining is a secondary process, by which stock is removed gradually from the rectangular blank to achieve the desired profile/shape, dimensional accuracy, and surface finish. Such processes can be broadly classified into conventional machining processes (i.e., turning, drilling, boring, shaping, planning, grooving, etc.), and abrasive machining processes (i.e., grinding, honing, polishing, etc.) (Stewart 1970).

These machining processes are subtractive in nature, where material removal occurs layer by layer, until the final profile or 3D features are obtained on the component/part. However, each of these machining processes have distinct characteristics in terms of stock removal rate, cutting speed, cost, and most importantly the resultant surface finish. In this context, most conventional machining processes are suited for high stock removal and providing acceptable surface finish. However, accomplishing both these objectives in a single pass is not conceivable. Consequently, machining operations is often performed in two or more steps, under different parameters (i.e., cutting speed, feed speed, and stock removal) (Aguilera et al. 2003).

In first rough machining operation, the stock removal rate is much higher. The higher stock removal rate is attributed to the higher feed rate and large depth of cut used. This step is called **rough cut or roughing pass**. Large sized chips and the high noise levels are prominent features of the rough machining operations.

To provide good surface finish and close tolerance, fine machining operations are undertaken. At low feed speeds with shallow depth of cuts, the stock removal rate is significantly reduced. Hence, surface quality improves, and so does dimensional accuracy and machining tolerances (Marchal et al. 2009). The chips produced are much smaller, usually a large proportion as fine wood dust, and the noise levels much lower in the fine machining operations (Table 5.1).

5.5 Material Removal Rate and the Process Parameters

Cutting velocity (V_c), feed rate (f) and depth of cut (t) are three process parameters for every conventional machining process. These parameters significantly impact the outcomes of the machining operation (Marchal et al. 2009). Higher velocity, feed, and depth of cut can increment material removal rate (MRR), but with a sacrifice of surface quality. MRR is determined by the cutting speed, feed, and depth of cut, and increasing any of this variable will have a positive impact on the stock removal. The cutting speed is a constant in most machining operations. It is pre-set depending on the nature of the workpiece, tool material, machine-tool interaction, vibration characteristics, as well as other related factors. In essence, higher feed and depth of cut are applied in rough machining operations to increase the MRR proportionately. The opposite is true for fine machining operations, which accomplishes it's objectives by lower stock removal rates (Csanády et al. 2019).

Table 5.1 Differences between rough and fine machining

Rough machining	Fine machining
The primary objective is to achieve high stock removal rate quickly	The primary objectives are to achieve low stock removal
Higher feed speed and depth of cut are used	Feed speed is low with small depth of cut
Higher chip load on knife/tool	Lower chip load on knife/tool
Large amount of material is removed	Small amount of material is removed
The resultant surface is poor and rough	The resultant surface finish is good and smooth
Dimensional accuracy is low	Dimensionally accurate, with tight tolerance
Worn out or old tools/knives can be used for this operation	Sharp tools/knives are necessary
It is the first operation performed	It can be performed after the rough machining operation
Large chip sizes and high noise levels are produced	Smaller chip sizes, usually a proportion as fine wood dust, and lower noise levels are produced

Adapted from Csanády et al. (2019)

5.6 Surface Finish and Dimensional Accuracy

The use of improper feed speeds often leads to the presence of cutter marks, or roller marks on the machined surface. Generally, the resultant surface roughness of machined wood is influenced not only by the cutting tool geometry but also the feed speed (Seki et al. 2013). Inevitably, higher feed speed usually impairs the resultant surface quality and dimensional accuracy of the machined workpiece. The opposite is true for lower feed speeds. However, lower feed speeds will impair the throughput rate, which is undesirable in furniture manufacturing, as it leads to increased production cost. On this account, lower feed speeds produce a higher number of cutter marks per unit length of the machined surface. To achieve high-quality machined surface, the pitch distance between successive cutter marks should be 1 mm or less, which necessitates an optimal throughput rate (Ratnasingam et al. 1999a).

Worn-out cutters (i.e., higher edge radius and nose radius) limits the quality of surface finish achievable in the machining process. However, worn-out cutters can be safely used in rough machining operations without noticeable changes, due to its high chip load capacity. On the other hand, sharp tools are necessary to achieve better surface finish, machining accuracy, and tolerance. The subject of woodworking tool wear, and its causes are well researched and reported in the papers by Klamecki (1979), Sheikh-Ahmad and Morita (2002), and Labidi et al. (2005).

From a production engineering perspective, finding a balance between the technical parameters of the machining operation and the process economics is important. One cannot be achieved without the other. Unfortunately, in many furniture manufacturing operations, this concept is often overlooked, often resulting in higher production cost. Since there are several operations involved in the machine shop, and to minimize the machine setup times, it is desirable to work out the production schedule in consultation with both the product and process engineering departments in the furniture factory. Such an approach will conceive the best production schedule for maximum economic benefit (Ratnasingam et al. 1999a).

It must be emphasized that the economics of the wood machining process (Csanády et al. 2019) is often increased by one, or all, of the following:

1. reducing the amount of stock removal,
2. machining with higher cutting speeds,
3. reducing the machining tolerances,
4. minimizing the amount of fine dust generated, and
5. encouraging recovery of off-cuts and reducing waste.

The quality of the machined surface in the machine shop is of primary concern, as it has a strong impact on the subsequent surface finishing operation. Abrasive sanding, laminating with veneer, direct printing, spray finishing, etc., are some of the subsequent operations, which dictates the degree of surface smoothness required. For furniture manufacturers however, the primary concern is finding the balance between production rate and cost. This balance is a function of the interaction between cutting forces, energy consumed, tool life, and resultant surface quality of the machining process. Optimal machining operations is derived from this balance and continue to be a subject of extensive research (Csanády and Magoss 2013).

Beyond the machine shop, the profiled and shaped components/parts are checked against the drawings and plans of the sub-assemblies and prototype pieces of the furniture, to ensure its conformance. The next step is assembling the furniture, before it is finished, although in some instances, the finishing operation would take precedence. In most instances, however, the furniture is assembled prior to finishing, to ensure the highest possible surface quality.

5.7 Panel-Based Furniture Manufacturing

Panel-based furniture, including office furniture, case-good, and kitchen cabinets has grown to become the second largest category of furniture types manufactured and exported throughout the world (CSIL 2019). Case-good is usually box-shaped with drawers and doors for storage. On the other hand, cabinets may have one or two doors with hardware, and sometimes with a lock. Cabinets may have drawers and/or shelves concealed by the doors. Low cabinets regularly have a finished surface top for display, or as a working surface, such as the countertops found in kitchens. Kitchen cabinets can be mounted on the wall or placed on the floor against the wall. Cabinets in bedroom are usually with one or more columns for clothing, attached with a few drawers, put one on top of the other. This is known as a ‘chest of drawers’ and is often used for storing small articles. Other types of cabinets found in the bedroom may include the bedside cabinet, a wardrobe, ‘armoire’. A tall cabinet for hanging of dress/shirts, is called a closet or an ‘armoire’, and a closet, usually refers to a built-in cabinet (Rae 2001).

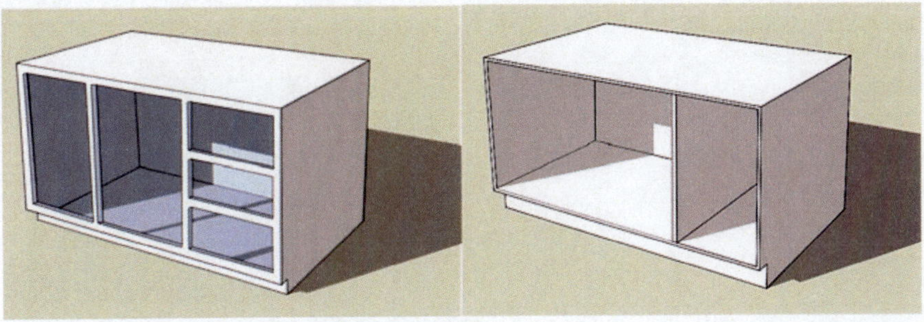

Fig. 5.5 Cabinet construction—frame and frame-less (Courtesy of the Malaysian Furniture Council)

There are two types of constructions used in cabinet manufacturing, i.e., framed or frame-less cabinet construction (i.e., European style-cabinet) (Natale 2009). Frame cabinets usually have a frame attached to the front of the cabinet box. Frame-less cabinet have no such supporting frame, and its doors directly attached to the sides of the cabinet box (Fig. 5.5). At present, most cabinets manufactured are frame-less, and are usually made of wood-based panels such as, *plywood*, *chipboard*, or *medium-density fibreboard*. These wood-based panels are usually over-laid with either low-pressure laminates (LPL), high-pressure laminates (HPL), melamine, thermo-foil, or wood-veneer, and in some instances, directly printed using coatings and stains. With such surfacing technologies, a variety of patterns and designs are available to the customers to choose from to suit their taste and needs (Tankut and Tankut 2010).

The 32 mm cabinet making system is a construction and manufacturing system used in the production of ready-to-assemble (RTA) and European-style frame-less cabinets, and other types of case-good furniture. The system is widely used throughout the world, partly owing to global furniture retailing giant 'IKEA', using some of its elements (principally the 32 mm shelf support holes) in its furniture constructions. The system facilitates standardization of both component's dimensions and production processes, and through custom designed fittings, makes the manufacturing process one of the most highly automated within the furniture industry (Kristoffersson 2014).

The manufacturing processes of wood panel-based furniture, such as cabinets, office furniture, etc., follows the steps shown in Figure 5.6.

The important machines used, and its respective functions in the manufacturing of wood panel-based furniture are shown in Figure 5.7.

Wood-based panel furniture manufacturing industry has the highest potential to adopt automation and advanced machining technologies compared to the other furniture sub-sectors (i.e., wooden, metal and leather/sofa). This is attributed to the fact that the wood-based panels, being an engineered wood composite, is less variable compared to solid wood material and the finished product contains a large proportion of components with straight-cut edges and shapes (Natale 2009). Wood-based panels furniture factories usually have a relatively simple process flow on the factory

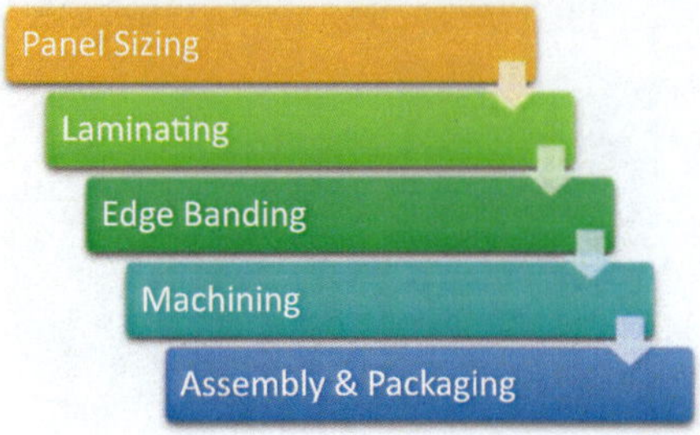

Fig. 5.6 Process flow for panel furniture manufacturing

shopfloor, with relatively short machining time and minimal laborious handling. In fact, it is the likely candidate for the application of Industry 4.0 technologies. In this context, wood-based panel furniture manufacturing industry has the highest productivity level among all other furniture manufacturing sub-sectors, due to its relatively higher use of technologies (Ratnasingam 2017).

Summary

- The rough milling operation aims to produce blanks/squares which the correct dimensions for further machining and free of defects.
- The rough milling operations involve higher stock removal rates at lower cutting speeds, which lead to large wood chips produced as waste.
- The machine shop imparts a geometric alteration to the blanks/squares, converting it into a profiles/shaped component with a fine finish.
- Due to the use of higher cutting speeds in the machine shop, the stock removal rate is decreased but finer wood dust is produced.
- In the case of wood-based panel furniture manufacturing, the process flow is relatively simpler, with the machine shop undertaking most of the value-adding operations until to furniture is produced.

<table>
<tr>
<td>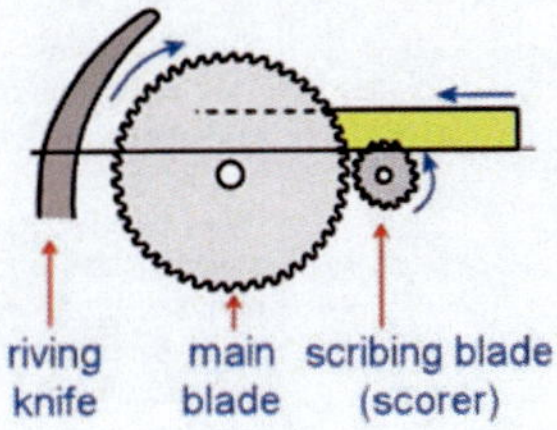
</td>
<td>Panel Sizing Saw
This machine cuts the wood-based panels to the required dimensions, in terms of length and width. In the latest machines, the best cutting plan is pre-determined by the computer to ensure the highest recovery/yield.</td>
</tr>
<tr>
<td>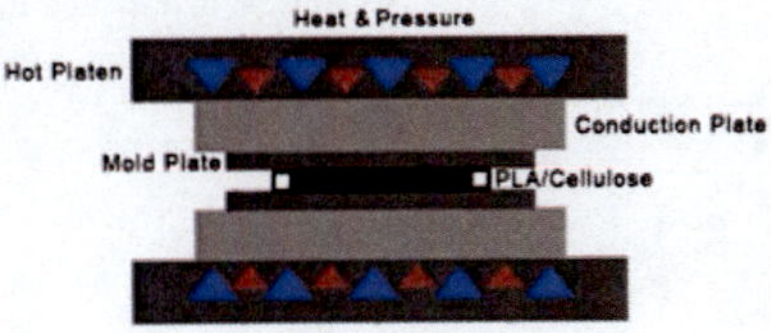
</td>
<td>Laminating Press
This operation involves the over-laying of the wood-based panels with specific type of laminates. If the panels are to be directly printed, this operation is by-passed. Direct printing is done using water-based coating material, which is applied using roller-coaters attached with conveyorized drying unit.</td>
</tr>
<tr>
<td>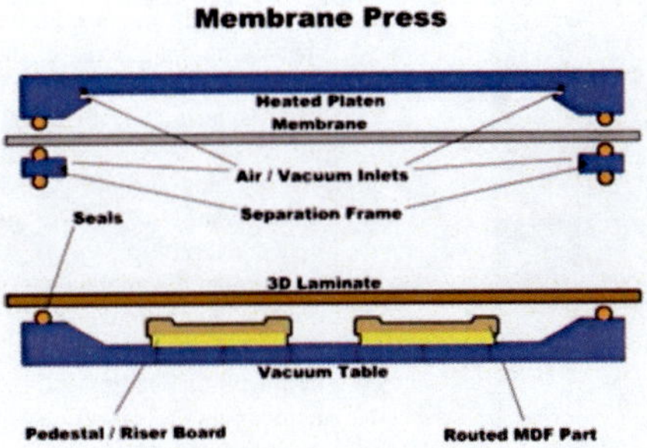

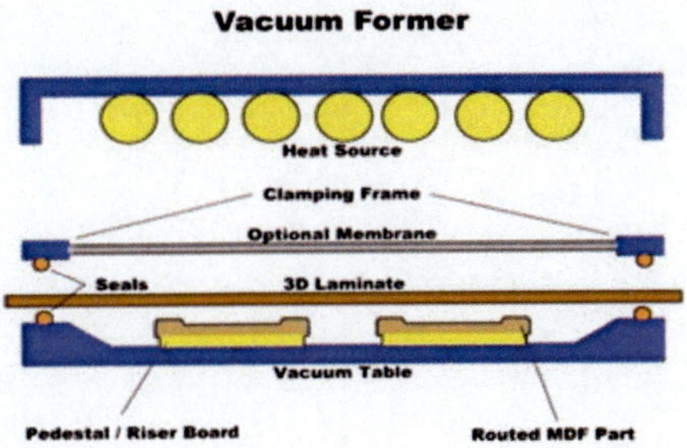
</td>
<td>Membrane Press
For laminating wood-based panels with 3D shapes or profiles, a membrane press or vacuum press is used. This equipment allows the application of paper or plastic laminates to be applied over the shapes/profiles to impart specific appearances.</td>
</tr>
</table>

Fig. 5.7 Common machines used in panel furniture manufacturing (Courtesy of the Malaysian Furniture Council)

<table>
<tr>
<td>
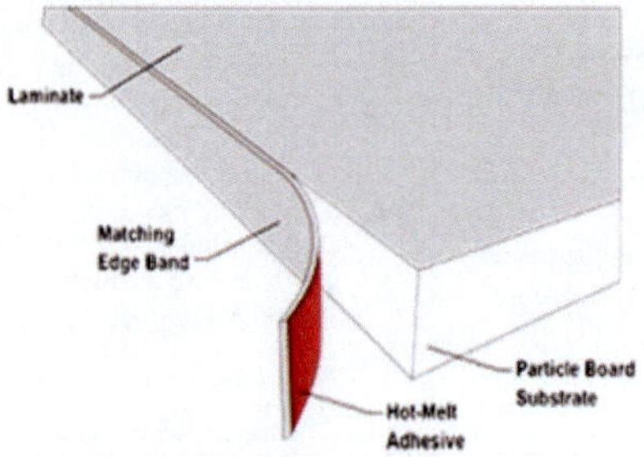

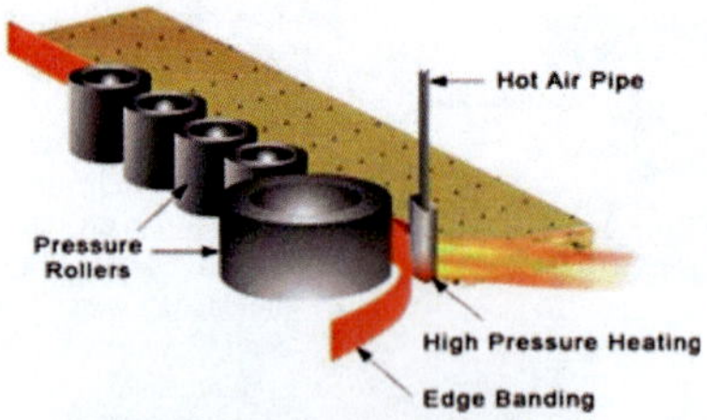

</td>
<td>
Edge-Banding
This operation involves the gluing of a thick plastic band on the edges of the wood-based panels. Usually, a hot-melt adhesive is used for this purpose. If the edges of the wood-based panels are profiled, a profile-wrapper, which operates on a similar technology as the edge bander, is used. The notable difference is that the thickness of the band is significantly less in profile wrapping.
</td>
</tr>
<tr>
<td>
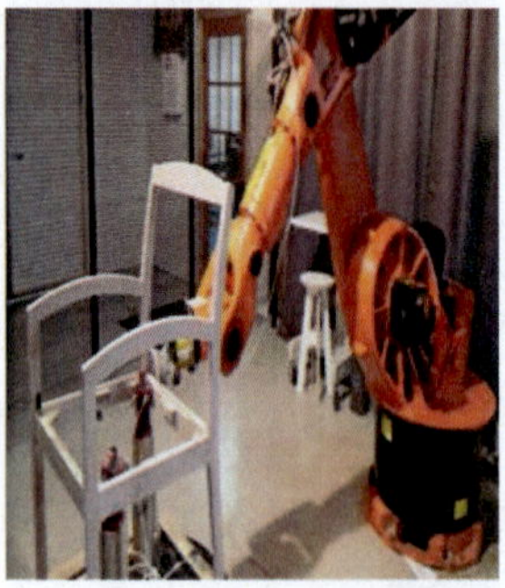
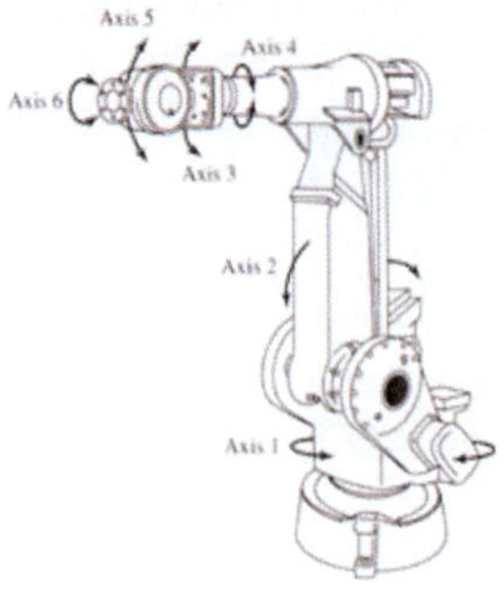

</td>
<td>
CNC Machining Centre and Robotic Arm
In advanced factories, all the machining operations for the wood-based panel furniture manufacturing is carried out using the CNC machining centers and robotic arms. With a range of tools in the magazine, multiple machining operations are undertaken accurately and rapidly. The robotic arms help with the loading and unloading of components/parts.
</td>
</tr>
</table>

Fig. 5.7 (continued)

References

Aguilera A, Méausoone PJ, Martin P (2000) Wood material influence in routing operations: the MDF case. Holz Roh Werkst 58(4):278–283

Aguilera A, Méausoone PJ, Martin P (2003) A new methodology for wood cutting optimization in the secondary manufacturing processes. Holz Roh Werkst 61(5):358–362

Bagnall R (2021) Beginner's guide to cnc machining in wood—understanding the machine, tools and software. Foxchapel Publishing, Mount Joy, Pennsylvania, USA

Conover E (2019) The lathe book—a complete guide to the machine and its accessories. Taunton Press, Newtown, Connecticut, USA

Collins JA, Busby H, Staab G (2010) Mechanical design of machine elements and machines—a failure preventive perspective. Wiley, Hoboken, New Jersey, USA

CSIL (2019) World furniture outlook. CSIL—Centre for Industrial Studies, Milan, Italy

Csanády E, Magoss E (2013) Mechanics of wood machining, 2nd edn. Springer International Publishing Cham, Switzerland

Csanády E, Magoss E, Tolvaj L (2015) Quality of machined wood surfaces. Springer International Publishing Cham, Switzerland

Csanády E, Kovács Z, Magoss E et al (2019) Optimum design and manufacture of wood products. Springer International Publishing, Cham, Switzerland

Davim JP (ed) (2013) Wood machining. Wiley, Hoboken, New Jersey

Jiang W (2019) Analysis and design of machine elements. CRC Press, Boca Raton, Florida, USA

Kilic M, Hiziroglu S, Burdurlu E (2006) Effect of machining on surface roughness of wood. Build Environ 41(8):1074–1078

Koch P (1964) Wood machining processes. The Ronald Press, New York, USA

Klamecki BE (1979) A review of wood cutting tool wear literature. Holz Roh Werkst 37:265–267

Koc KH, Erdinler ES, Hazir E et al (2017) Effect of CNC application parameters on wooden surface quality. Measurement 107:12–18

Kristoffersson S (2014) Design by IKEA. Bloomsbury Publishing Inc., Broadway, New York, USA

Labidi C, Collet R, Nouveau C et al (2005) Surface treatments of tools used in industrial wood machining. Surf Coat Tech 200(1–4):118–122

Mohammadi S, Mirzapour SMJ, Rekik Y (2020) An integrated production scheduling and delivery route planning with multi-purpose machines: a case study from a furniture manufacturing company. Int J Prod Econ 219:347–359

Marchal R, Mothe F, Denaud LE et al (2009) Cutting forces in wood machining—basics and applications in industrial processes. A review COST Action E35 2004–2008: Wood machining-micromechanics and fracture. Holzforschung 63(2):157–167

Merdzhanov V (2018) Optimization of technological parameters for continuous edge-banding of furniture panels. Annals of Warsaw University of Life Sciences No. 102—Forestry & Wood Technology 112–119

Natale C (2009) Furniture design and construction for interior designers. Fairchild Publication, New York, USA

Rae A (2001) The complete illustrated guide to furniture and cabinet construction. Taunton Press, Newtown, Connecticut, USA

Ratnasingam J (2017) Automation and advanced technology in the ASEAN furniture industry. Tech. Note No. 11, IFRG Publication, Singapore

Ratnasingam J, Scholz F (2007) Characterizing surface defects in machine-planing of Rubberwood (Hevea brasiliensis). Holz Roh Werkst 65(4):325–327

Ratnasingam J, Scholz F (2008) Yield studies of Rubberwood lumber during rough milling operations. Holz Roh Werkst 66(6):467–468

Ratnasingam J, Ma TP, Perkins MC (1999) Productivity in wood machining processes-a question of simple economics. Holz Roh Werkst 57(1):51–56

Ratnasingam J, Reid HF, Perkins MC (1999) The productivity imperatives of coated abrasives application in furniture manufacturing. Holz Roh Werkst 57:117–120

Simanová L, Gejdoš P (2015) The use of statistical quality control tools to quality improvements in the furniture business. Procedia Econ Financ 34:276–283

Seki M, Sugimoto H, Miki T et al (2013) Wood friction characteristics during exposure to high pressure: influence of wood/metal tool surface finishing conditions. J Wood Sci 59:10–16

Sheikh-Ahmad JY, Morita T (2002) Tool coatings for wood machining: problems and prospects. Forest Prod J 52(10):43–47

Stewart HA (1970) Abrasives versus Knife Planing. Forest Prod J 20(7):43–47

Tankut AN, Tankut N (2010) Evaluation of effects of edge banding type and thickness on the strength of corner joints in case-type furniture. Mater Design 31(6):2956–2963

Chapter 6
Joints in Furniture

The furniture parts and components produced are assembled into sub-assemblies, or furniture, using joints. There are several different types of furniture joints, but the most common are the mortise-tenon and dowel joints. Joints must be machined with precision and accuracy to ensure a tight fit, which produces a strong joint. The machining tolerance allowed is small, to result in a thin glue-line thickness, which facilitates strong joint formation. Many different glue or adhesive types are available in the market for joint making, and the suitable choice for the application must be selected. This chapter provides an overview of the furniture joint types, its machining, and its gluing operations.

6.1 Introduction

Joints allow the assembly of different components to form the furniture structure or frame, which is not only aesthetically appealing, but also structurally sound. Wood, being a natural and variable material, is also hygroscopic. Inevitably, it is subject to movements caused by the changes in relative humidity in the environment. Although these movements are relatively small, and predictable in most instances, it has a profound effect on joint design (Rae 2001).

Joints are the weakest part of any furniture, and they are the most important cause of failure in furniture. In the design and specification of furniture, the stresses acting on the joints must be considered to ensure the strength and durability of the structure. Since every joint type differs in its design and development features, its strength characteristics must be known, especially its performance when subjected to the various forms of loads, such as bending, shear, tensile, and fatigue, by carrying out the necessary strength testing (Eckelman 1978). In many furniture constructions, various glues/adhesives, and fasteners (i.e., nails, screws, bolts) are often used to enhance the structural strength as well as the effectiveness and rigidity of the joints.

J. Ratnasingam, *Furniture Manufacturing*, Design Science and Innovation,
https://doi.org/10.1007/978-981-16-9412-7_6

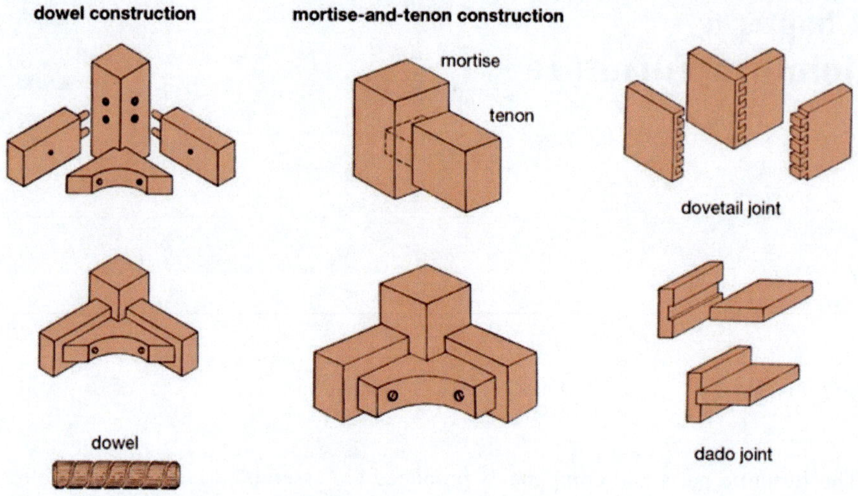

Fig. 6.1 Common furniture joints (Courtesy of the Malaysian Furniture Council)

There are many types of furniture joints, and they differ in ease of production, inherent strength, and intended application (Fig. 6.1).

There are different joint types used in furniture, and the choice of joint used is often determined by the ease of manufacturing, strength performance, design suitability, and cost (Eckelman 1978; Spagnuolo 2020). Among the common furniture joint types include the (1) **miter joint**, which is made when two end pieces are cut on angles of 45° and fitted together. It is commonly found within the corners of a structure; (2) **butt joint** is where the square end of one piece butts into the side or the end of the other piece and often fastened by nails or screws; (3) **lap joint** is simply where two pieces of wood overlap, often to extend its length, (4) **tongue and groove joint** is applied in flat and parallel pieces of wood, and features a tongue, or edge, running along one side and an indented groove running along the other side; (5) **rabbet and dado joint** is formed when a piece of wood called a rabbet is inserted into a groove, the dado, and reinforced with glue; (6) ***finger joint*** is made by cutting a set of complementary, interlocking profiles of fingers in two pieces of wood, which are attached by using adhesives; (7) **dovetail joint** is an exceptionally solid joint, and can withstand pulling through the use of wedge-shaped interlocking pieces (the wedges take after a dove's tail) regularly found where two pieces of wood meet at a right angle, such as along the corners of drawer sides; (8) **mortise and tenon joint** is formed by shaping one end of the wood to be embedded into the cavity in another piece of wood. The mortise is the cavity, while the tenon is the shaped piece that fits into the mortise; (9) **dowel joint** is comparable to the mortise and tenon, in which a round peg is fitted into a cavity/bore hole to fortify a joint. The distinction is that a dowel is more often than not a shaped round, cylindrical peg, where both ends are inserted into cavities of the jointed pieces of wood (Rae 2001).

Fig. 6.2 Mortise and tenon joint

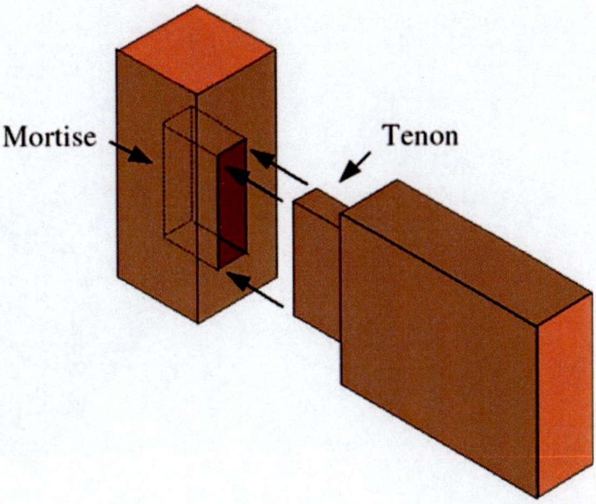

In the strength design of furniture, joints are the weakest link in the structure, and, as a result, joint strength has a strong impact on the performance of furniture during service. In this respect, further deliberations on the common furniture joint types are provided below, although extensive technical description on this subject is provided in Eckelman (1978). This is an important point as furniture in service is subjected to different load-bearing capacity during its life.

One of the most common furniture joints is the mortise and tenon joint. The mortise is rectangular cavity, which precisely fits the rectangular projection, known as the tenon (Fig. 6.2). Due to its higher bending moment, such joints are much stronger than the dowl joint and have higher load-bearing capacity (Csanády et al. 2019).

Dovetail joint is a frame of box joint, where diagonally cut fingers are locked to produce very strong joints. These types of joints have long been used in constructing drawers and cabinets (Fig. 6.3), but today their application is limited due to higher production cost and the more work required (Csanády et al. 2019).

Finger joint is another form of box joint, in which the ends of the pieces of wood are locked by interlocking fingers, and the interlocking fingers receive pressure from two directions (Fig. 6.4). The developments in finger jointing have revolutionized the woodworking industry, as it has enabled the recovery of short pieces of wood, and off-cuts, which were previously discarded as waste (Ratnasingam & Scholz 2009). With the correct adhesives finger joints are relatively strong and can be used for many structural applications (Eckelman 1978; Rae 2001). Finger jointed boards or panels of large dimensions have found application as tabletops, headboards, etc.

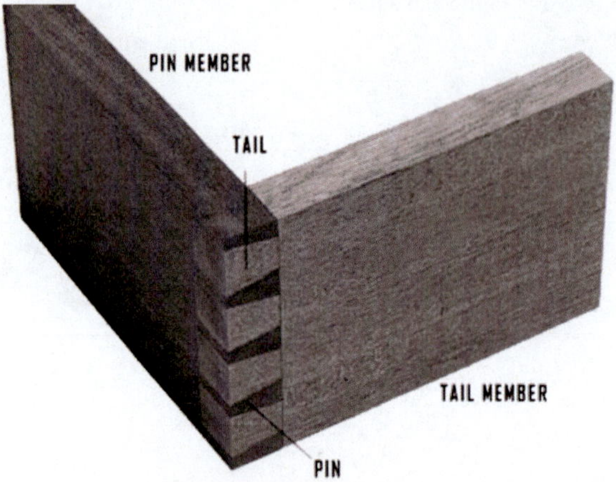

Fig. 6.3 Dovetail joint

Fig. 6.4 Finger joint

The most widely used joint in furniture is the dowel joints. Dowels are cylindrical wooden pegs, which are inserted and glued into another piece of wood (Fig. 6.5). They are cheap, produce reasonably strong joint, and can be manufactured quickly in the production line of furniture factories (Ratnasingam and Ioras 2013, 2015). Some dowels have plain surface, while others have serrated surface to increase net effective surface area for glue/adhesive application (Csanády et al. 2019). The latter has higher load-bearing capacity as compared to the former due to its higher effective area.

Fig. 6.5 Dowel joint (Courtesy of the Malaysian Furniture Council)

6.2 Machining of Furniture Joints

The machining of these different furniture joints is carried out using specific machines, such as the boring machine, tenoner, mortiser, or general machines, such as a shaper with special profiled tools to machine the different types of joints (Fig. 6.6). A good description of woodworking cutting tools is provided in Effner (1992), Lee (1995), and Hall (2017). In recent years, however, the CNC workstations, with its combination of different cutting tools, have been able to machine these joints rather quickly. Joints must be machined to have a tight fit, and hence it must be accurately and precisely machined with tolerances of about 0.15 mm.

Machined furniture joints must have a close fit, usually with allowances of less than 0.15 mm. If the tolerances are much larger, it is overcome with a thicker glue line, which is not desirable as it weakens the joints significantly. Ideally, joints with tight fits are highly desirable as they transfer the load from one member to another effectively, within the furniture structure (Eckelman 1978).

6.3 Adhesives for Furniture Joints

Furniture joints are usually strengthened using glues/adhesives, or fasteners, which reduces any movements in the joints when subjected to load. There are several types of glues/adhesives commonly used in the furniture industry, and the criteria for selecting the choice of glue/adhesive are (1) ease of application, (2) ease of curing, (3) strength performance, and (4) cost. A detailed discussion of the different types of furniture adhesives, its chemistry, and application is provided in Tout (2000), Sellers (2001), and Hunt et al. (2018).

Type of Machine	Function
 	**Boring Machine** It produces smooth and exact bore holes in a workpiece. The boring bit used, bear a single cutting tip of steel, cemented carbide, or precious stone. Beyond the bit, the drills have flutes, of specific diameter and length. Tool vibrations during boring must be eliminated avoid bore-hole variations.
 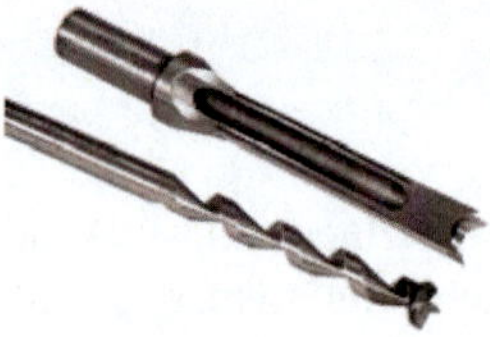	**Mortiser** Used for producing round or square holed mortises, with the slot produced from overlapping the round or square holes. Similar in action to a pedestal drill, with a lever pulling down the cutter to execute the machining.
	Tenoner Used to cut tenon profiles for the mortise holes, across the grain. The workpiece is held in place by pads.

Fig. 6.6 Common machines for producing furniture joints (Courtesy of the Malaysian Furniture Council)

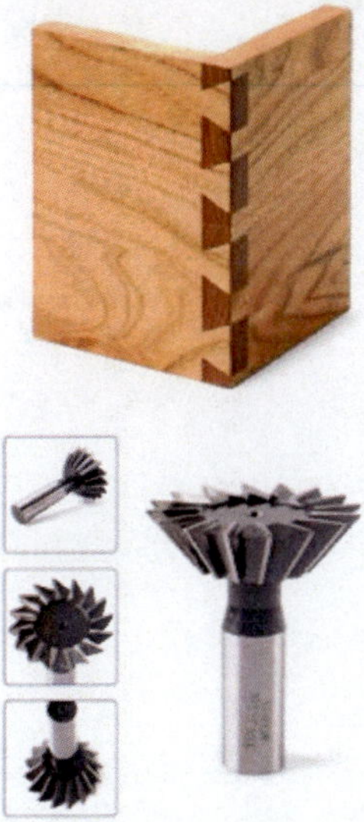	**Dove-Tail Cutter** Dovetail cutters are specialized tools for cutting dovetail joint. They may be of solid profile, or with a holder and inserts. The cutter is attached to a horizontal spindle machine.

Fig. 6.6 (continued)

Polyvinyl acetate (PVA) (also known as white glue) is the most common glue/adhesive in furniture manufacturing. It is non-toxic and easy to apply, but difficult to repair since most glues do not adhere well to solidified PVA glue. It is a type of thermoplastic emulsion glue and will creep under constant load (Bandel 1995).

Polyurethane (PU) glue is gaining popularity in the furniture industry. The polyurethane reactive (PUR) hot melt glue gives a very strong bond yet flexible and can be applied on difficult to bond wood materials, especially those with extractives (Hse and Kuo 1988).

 	**Finger Cutter** It is a constant profile tool, with where the tool body thickness corresponds to finger pitch. Finger joints may be with, or without shoulders. The finger cutters are often attached to spindle molder or shaper.
 	**Miter Cutter** The miter saw produces a crosscut at a desired angle on a workpiece, using a saw that rotates in a spinning direction.
	**Rabbet and Dado Cutter** This is a type of matched cutters or complementary cutters, often used on a spindle molder or router, to machine precisely matched joints.

Fig. 6.6 (continued)

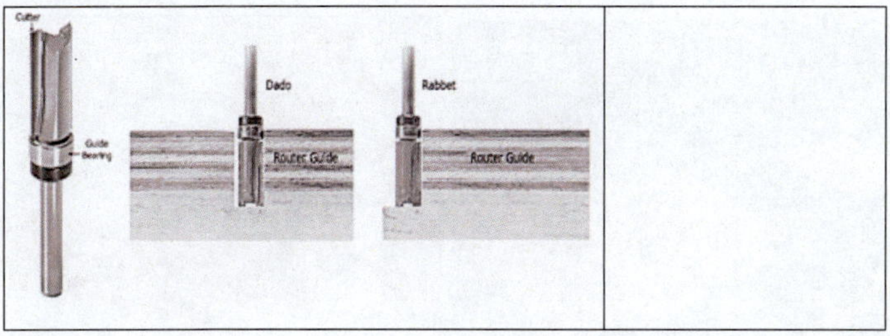

Fig. 6.6 (continued)

Epoxy glue is a two-part adhesive system, which cures under a wide range of temperatures and humidity. It does not require pressure to cure and has great gap-filling properties. Despite its superior strength, epoxy glues are cumbersome in application.

Cyanoacrylate (CA) glue (also known as super-glue) is common for minor repairs in the furniture industry. It bonds immediately, and the cured CA glue is basically a plastic material, with good gap-filling ability (Frihart 2015).

Aliphatic Resin (AR) (also known as carpenter's glue or yellow glue) is a light yellow color synthetic-adhesive used to bond together pieces of wood. It has low odor and combustibility, but average in bonding strength moisture resistance. Nevertheless, it is still superior to the polyvinyl acetate glues. AR glue lines are however hard and cannot be finished over (Bandel 1995).

Animal Glue or commonly referred to as hide glue is presently used for the manufacture of musical instruments and replica furniture.

Thermoplastic Hot Melt (THM) is using ethyl vinyl acetates (EVA), and cures by cooling. Fast cure rate and easy handling are their main advantages. It is often used to tack a joint together in furniture prototypes before the main adhesive has cured.

6.4 Adhesive Curing

In building the furniture sub-assembly or frame, the components/parts jointed must be held in position until it achieves the necessary strength for further handling and processing. In doing so, the joints are usually tacked and held in position using a clamp or clamping machines, until the glues/adhesives applied cures (Fig. 6.7).

The furniture joints must be machined accurately and precisely to achieve a tight fit, allowing only 0.05–0.15 mm gap in between the two surfaces. Such a tolerance is a necessity to ensure optimum glue-line thickness for sufficient bond strength. The clamping process usually applies approximately 2 N/mm^2 or 2 bars of pressure to ensure close contact between the two surfaces to facilitate a good bond formation (Bandel 1995). It has also been shown that when higher temperatures (i.e., between

Fig. 6.7 Clamping techniques for joints (Courtesy of the Malaysian Furniture Council)

45 °C and 80 °C) are used, the curing and hardening of the glue is accelerated, with the joint attaining full bond strength faster, than when it is left to harden in ambient conditions. The higher temperatures can be achieved through the application of dry saturated steam or radio-frequency (RF) heating (Fig. 6.8).

Since the joints are the weakest link in the furniture structure (Eckelman 1978), efforts are usually taken to ensure the joint design and its construction is fortified using the correct glue/adhesive and/or fasteners (Fig. 6.9). Fasteners are devices that mechanically join, or affix two or more components/parts, or sub-assemblies together. In general, fasteners do not form permanent joints, i.e., joints that can be assembled and dissembled without any damage to the jointing components (Csanády et al. 2019).

One important innovation in furniture jointing is the advent of the Threespine™ patented click technology that has transformed furniture jointing completely, especially for wood-based panel furniture. Furniture can now be assembled and disassembled in seconds without using any tools, screws, fasteners, or other loose parts. It simply uses a plastic tongue and groove mechanism to hold furniture parts together. The technology provides a perfect fit with stable results (Anon 2021).

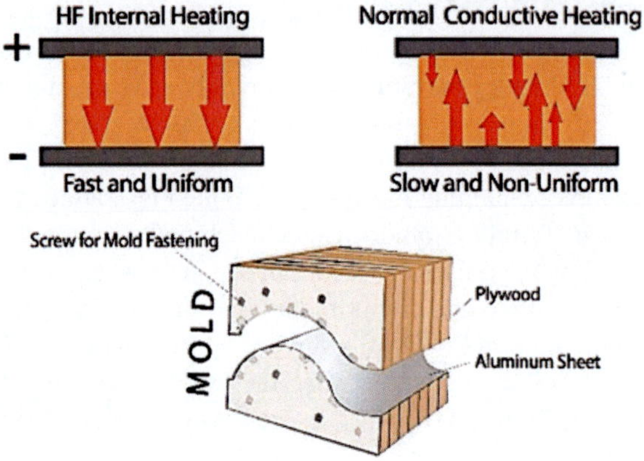

Fig. 6.8 Radio-frequency technology for adhesive curing (Courtesy of the Malaysian Furniture Council)

Fig. 6.9 Common furniture fasteners/fittings (Courtesy of the Malaysian Furniture Council)

6.5 Criteria For Selecting Adhesives

In selecting the correct adhesive/glue for furniture, several criteria must be taken into consideration. Although it includes strength performance, ease of application, curing requirements, cost, and environmental compliance, the customers may also impose their will on the choice of glue/adhesive to be used. For instance, furniture to be exported into the United States of America market must comply with the California Air Resources Board (CARB) formaldehyde emission standard (CSIL 2019), and therefore customer will insist on glues/adhesives that meet this requirement in their furniture. Although the strength performance of glues/adhesives is of priority (Stoeckel et al. 2013), the other criteria must also be taken in consideration, before dicing on the choice of glue/adhesive for a particular furniture application. Detailed and elaborate discussions on the chemistry, application, and performance of the various glues/adhesives applied in the woodworking industry are provided in the reports by Marra (1992), Pizzi & Mittal (1994), and Bandel (1995).

1. **Application Requirements of the Glue/Adhesive**
   - Degree of surface preparation.
   - Tackiness.
   - Cure conditions.
   - Viscosity—pseudoplastic and thixotropic qualities are desirable.
   - Application and system cleaning ease.
2. **Physical Properties of the Cured Glue/Adhesive**
   - High cohesive strength.
   - Flexibility improves peel strength.
   - Strong bond line.
   - High damping capacity of the adhesive to disseminate vibration stresses, which improves the overall fatigue strength.

3. **Environmental and Chemical Resistance of the Cured Glue/Adhesive**

- Odorless and free of toxic emissions.
- Withstand environmental weathering (moisture, heat, UV light, etc.).
- Resistance to household chemicals and liquids.

In a recent publication by Kumar and Pizzi (2019), the environmental perspective of glues/adhesives used in the woodworking industry is elaborated in depth. The above considerations are becoming increasingly important as sustainable and environment-friendly furniture manufacturing practices are gaining traction throughout the global furniture market (CSIL 2019).

Summary

- Joints are the weakest links in the furniture structure.
- Joints must be machined with minimal tolerances, for a tight fit.
- The glue line in the joint must be of optimal thickness to effectively transfer the load from one member to another.
- Glues and fasteners are used to fortify the joint strength, but it cannot overcome a poorly designed joint.

References

Anon (2021) *Threespine* furniture click technology. Välinge Innovation, Viken, Sweden

Bandel A (1995) Gluing wood. CATAS Publication, Udine, Italy

CSIL (2019) World furniture outlook. CSIL—Centre for Industrial Studies, Milan, Italy

Csanády E, Kovács Z, Magoss E et al (2019) Optimum design and manufacture of wood products. Springer International Publishing, Cham, Switzerland

Effner J (1992) Chisels on a wheel—a comprehensive reference to modern woodworking tools and materials. Prakken Publication, Ann Arbor, Michigan, USA

Frihart CR (2015) Introduction to wood adhesives: past, present, and future. Forest Prod J 65(1–2):4–8

Hall H (2017) Tool & cutter Sharpening. Fox Chapel Publishing, Mount Joy, Pennsylvania, USA

Hse CY, Kuo ML (1988) Influence of extractives on wood gluing and finishing-a review. Forest Prod J 38(1):52–56

Hunt CG, Frihart CR, Dunky M et al (2018) Understanding wood bonds–going beyond what meets the eye: a critical review. Rev of Adhes Adhes 6(4):369–440

Kumar RN, Pizzi A (2019) Adhesives for wood and lignocelulosic materials. Wiley, Hoboken, New Jersey, USA

Lee L (1995) The complete guide to sharpening. The Taunton Press, Newtown, Connecticut, USA

Marra AA (1992) Technology of wood bonding—principles in practice. Van Nostrand Reinhold, Fifth Avenue, New York, USA

Pizzi A, Mittal KL (1994) Handbook of adhesive technology. Marcel Dekker Inc., Madison Avenue, New York, USA

Rae A (2001) The complete illustrated guide to furniture and cabinet construction. Taunton Press, Newtown, Connecticut, USA

Ratnasingam J, Ioras F (2013) Effect of adhesive type and glue-line thickness on the fatigue strength of mortise and tenon furniture joints. Eur J Wood Wood Prod 71(6):819–821

Ratnasingam J, Ioras F (2015) The fatigue characteristics of two-pin moment resisting dowel furniture joints with different assembly time and glue-line thickness. Eur J Wood Wood Prod 73(2):279–281

Ratnasingam J, Scholz F (2009) Optimization of finger-jointing in Rubberwood processing. Holz Roh Werkst 67(2):241–242

Sellers T Jr (2001) Wood adhesive innovations and applications in North America. Forest Prod J 51(6):12–16

Spagnuolo M (2020) Essential joinery. Blue Hills Press, Whites Creek, Tennessee, USA

Stoeckel F, Konnerth J, Gindl-Altmutter W (2013) Mechanical properties of adhesives for bonding wood—a review. Int J Adhes Adhes 45:32–41

Tout R (2000) A review of adhesives for furniture. Int J Adhes Adhes 20(4):269–272

Chapter 7
Sanding Process

The sanding process is the process of smoothening and flattening the surface of the workpiece using coated abrasive. It is a necessary operation not only to remove surface blemishes, but also to prepare the surface for the subsequent application of finishes or coatings. Unlike conventional machining operation, sanding involves a scrapping action, and thus a series of scratching operations are done using abrasives of different sizes, until a surface flat and smooth to the feel is obtained. Machine and hand sanding are commonly applied, and it is a process that generates high amounts of fine dust. This chapter provides an in-depth discussion of the sanding process, the machines involved, and the process outcomes.

7.1 Introduction

There are two types of abrasives used in the woodworking industry throughout the world, namely, bonded abrasives and coated abrasives. The former refers to grinding tools used to sharpen knifes and cutters used in conventional machining processes (Malkin 1989), while the latter refers to abrasives coated on different mediums, used primarily to smoothen the surface of the workpiece.

The abrasive sanding process is one of the important value-addition operations in furniture manufacturing (Csanády et al. 2019). It lays the foundation for the machined surface by making it flat and smooth, to receive the finish or coating. Therefore, abrasive sanding is a pre-requisite for all machined components and sub-assemblies to achieve high-quality finish (Ratnasingam et al. 1999).

Wood materials in its natural form are textured, depending on its anatomical characteristics (Fig. 7.1). The fine textured wood tends to produce relatively smoother surface compared to the coarse texture wood. In this context, roughness in wood materials is attributed to anatomical roughness and the machining roughness, due to the various machining operations (Murmanis et al. 1986). Hence, the primary objective of the abrasive sanding process is to overcome this roughness on the wood

J. Ratnasingam, *Furniture Manufacturing*, Design Science and Innovation,
https://doi.org/10.1007/978-981-16-9412-7_7

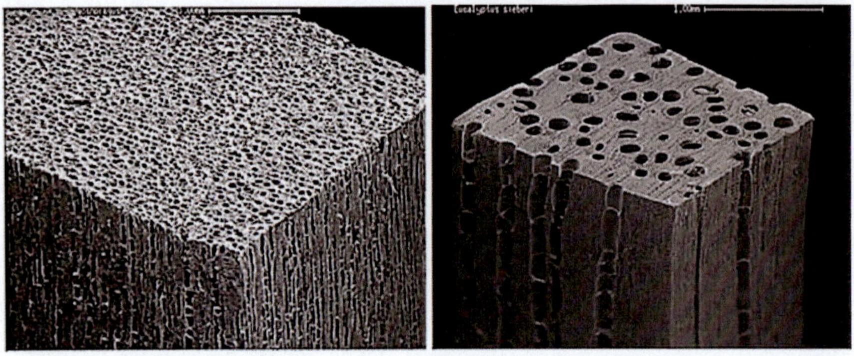

Fig. 7.1 Fine and coarse texture of wood

surface, by producing a smoot hand flat surface (Ratnasingam et al. 1999; Sofuoğlu and Kurtoğlu 2015).

The abrasive sanding process is also the last machining operation in furniture manufacturing and is carried out using different sanding machines. In theory, sanding is categorized as an orthogonal cutting process, due to the randomly distributed spherical cutting edges with negative rake angles, which removes material by crushing, scratching, and scrapping (Csanády et al. 2019).

In furniture manufacturing, abrasive sanding removes the top layer of the workpiece (usually less than 0.25 mm in thickness), ensuring a smooth and uniform surface. The aim is also to remove surface defects arising from previous operations, such as scratches, glue marks, etc. Abrasive sanding is also carried out after applying the base coat and before applying the finish coat to eliminate any raised grains and to improve the inter-coat adhesion bond of the layers of coating (Ratnasingam et al. 2004; Hendarto et al. 2006).

In general, the two main objectives of the abrasive sanding process are (1) thickness calibration (i.e., flattening the surface) and (2) smoothing of the surfaces to the desired degree of roughness to facilitate gluing or surface finishing. Due to the spherical/rounded profile of mineral grits, which act as the cutting edges, the upper surface layer of the workpiece is always crushed to a given depth depending on the size of the mineral grit and the pressure applied on the mineral grits. Due to the crushing effect and the resultant clogging at the surface of the workpiece, the measured surface roughness may be lower than the inherent roughness of the workpiece attributed to its anatomy (Luo et al. 2014; Csanády et al. 2015; Kúdela et al. 2018).

The crushing and scrapping effect of the abrasive sanding process is characterized by its ineffectiveness in energy consumption as much of the energy is lost as frictional heat, and the surface will have to be gradually smoothened by a series of smaller sized mineral grits (Porankiewicz et al. 2010). Inevitably, it generates fine and respirable wood dust, which poses risk to workers health and safety (Očkajová et al. 2018; Ratnasingam et al. 2011, 2019).

7.2 Structure of Coated Abrasives

The abrasive sanding, as the name implies, is undertaken by small sharp abrasive particles or mineral grits that are attached to a strong backing material (Fig. 7.2). As a result of the irregular distribution of the mineral grits, the abrasive action is often done by crushing and scrapping, which in turn results in high frictional forces. Therefore, the energy required for material removal is high, and a substantial amount of heat is produced during the process (Carrano et al. 2004; Ratnasingam et al. 2004). Due to the continuous wear of the abrasive mineral grits, the effective stock removal rate of the abrasive decreases over time. It must be emphasized that the rate of abrasive wear is a function of several parameters, including workpiece characteristics, and the nature of the sandpaper (i.e., mineral grit material, backing type, bond, open or close coated) which has been well researched and reported by Ratnasingam et al. (1999), Ratnasingam et al. (2004), and Csanády et al. (2015). The different types of minerals used as abrasives and the structure of an abrasive tool are shown in Fig. 7.3.

The most common abrasive minerals used in the furniture industry are aluminum oxide, garnet, silicon carbide, and ceramic. Generally, aluminum oxide grains create long-lasting and evenly cutting sandpaper, while garnet produces a softer cutting action, which inevitably also accelerates its wear. Silicon carbide produces a much better surface smoothness than aluminum oxide at the coarsest mineral grit sizes, for all wood species (Ratnasingam and Scholz 2008). Another advantage of silicon carbide is its higher heat conduction coefficient, which imparts superior performance during sanding of harder and abrasive workpieces, such as wood-based panels, including particleboard, medium density fiberboard, etc. This is because the high

Fig. 7.2 Sand paper (Courtesy of the Malaysian Furniture Council)

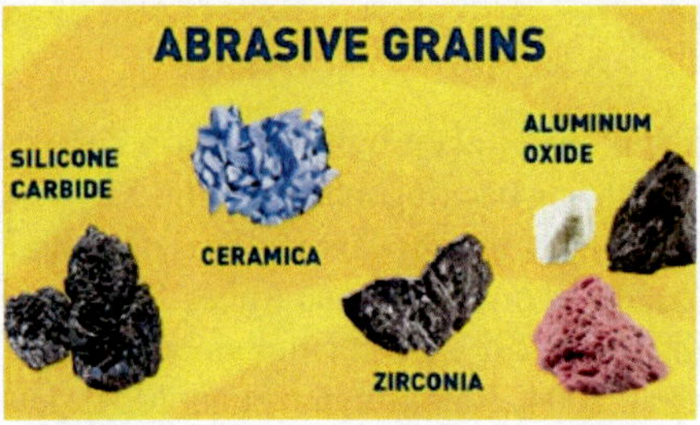

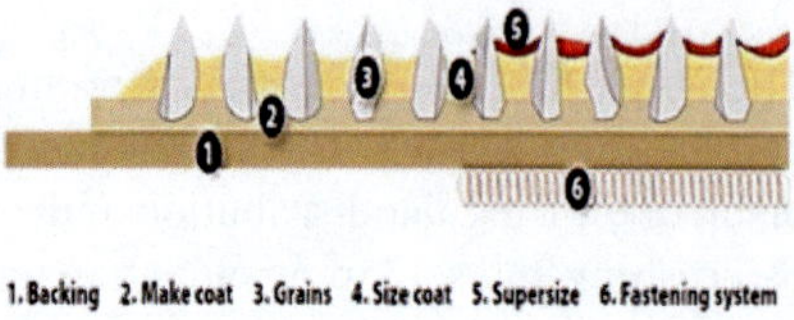

Fig. 7.3 Types of minerals and the structure of coated abrasive

adhesive and impurities content in these wood-based panels make its abrasive sanding very demanding (Ratnasingam et al. 2005). Nevertheless, the use of aluminum oxide abrasives is generally accepted and well established for abrasive sanding of furniture (Ratnasingam et al. 2004).

The common backing materials for coated abrasives are paper, film, and cloth (i.e., *cotton*, *polyester*, *rayon*). Paper is a flexible backing material, which allows irregular contours of a workpiece to be handled easily. Within the paper backing, there are several grades of papers, differentiated by its grammage/weight, designated by the alphabet A to F, in the order of reducing flexibility. Film backing is the most expensive due to the high degree of flexibility and is used for high-tolerance abrasive sanding operations, such as in optical instruments industry. It is usually made up of thin polyester films. On the other hand, relatively inflexible backing (i.e., cloth) is ideal for flat and level surfaces. The different grades of cloth backing available in the market are designated by the alphabets J, X, and Y, in the order of increasing stiffness. The harder the backing material, the faster is the sanding, which increases abrasive wear, leading to higher resultant surface roughness. The use of paper as the most common backing material explains the wide use of the general term of sandpaper to describe abrasive tools (Nagyszalanczy 1997).

The types of adhesives used in the manufacture of coated abrasives include hide glue (meant for general use only) and resin bond (meant for heavy-duty application). The former has lower heat resistance, while the latter can withstand higher operational heat (Ratnasingam et al. 2004).

As a rule of thumb, open-coated sandpaper (i.e., the abrasive grains cover between 50% and 70% of the surface) is better suited for soft woods, while close-coated (i.e., up to 95% of the surface is filled with abrasive grains) sandpaper is meant for hard woods. Since softwood produces more fibers as opposed to hardwood, it tends to clog up the sandpaper faster, thereby reducing the effectiveness of the abrasive sanding operation (Ratnasingam et al. 2004). In some sandpapers, anti-loading coatings are applied to reduce clogging.

The mineral grit size (or rather the size of the abrasive grains) is very important feature of the sandpaper, as it determines the rate of stock removal. Mineral grit sizes are variable and can range from P40 to P400 for furniture applications. This reflects the diameter of the individual mineral grit between 400 and 36 μm (Ratnasingam

Table 7.1 Abrasive grit comparison

Macro-grits		Micro-grits	
Grit	Mean diameter of mineral grain in μm	Grit	Mean diameter of mineral grain in μm
P60	269	P240	58
P80	201	P280	52
P120	125	P320	46
P150	100	P360	40
P180	82	P400	35
P220	68	P600	30

et al. 2004). In other words, lower grit numbers (P) correspond to coarser mineral grains and vice versa. For calibration purposes (i.e., large stock removal), the coarser grits (P40 and P60) are commonly used. For surface preparation (smoothening), the desired level of surface smoothness will determine the mineral grit sizes used. Should the starting surface be marred with obvious roughness, a sequence of increasing grit numbers will be used, to achieve the desired level of surface smoothness on the workpiece (Ratnasingam and Scholz 2004). In finishing operations, intermediate sanding is frequently required to facilitate the application of a new layer of coating, or surface finish, and for such applications finer mineral grit sizes between P300 and P400 are commonly used (Table 7.1).

The mechanics of the abrasive sanding process is complex, as it involves analyzing the true shape of the mineral grains, which are arbitrarily dispersed on the sandpaper (Luo et al. 2020). Especially when working with variable, non-homogeneous material, such as wood and wood-based panels, the abrasive sanding process is more complicated because other factors must be taken into consideration (Csanády et al. 2015, 2019). Nevertheless, with some simplifications, it has been possible to establish relationships describing the abrasive sanding process as a function of cutting and feed speeds, sanding pressure, contact length, and wood species (Ratnasingam et al. 2004). These relationships although derived experimentally have been useful to explain variations in the stock removal rate as a function of the working time. The kinematics of the abrasive sanding process as depicted by a single abrasive mineral grain is shown in Fig. 7.4.

However, the differences between conventional machining and abrasive sanding must be well understood (Fig. 7.5). Apart from eliminating or minimizing the natural anatomical roughness and surface wood damage, associated with machine cutting, the abrasive sanding process must ensure that it does not impart scratch marks more than necessary on the surfaces of the workpiece (Ratnasingam et al. 2002). This is particularly important since there is no minimum amount of stock removal standard to achieve a smooth surface, and it is a subject of production economics.

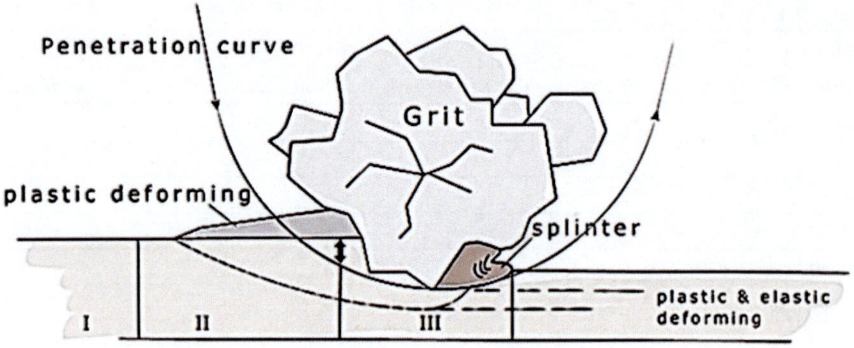

Fig. 7.4 Action of an abrasive grain

Fig. 7.5 Difference between conventional machining and abrasive sanding

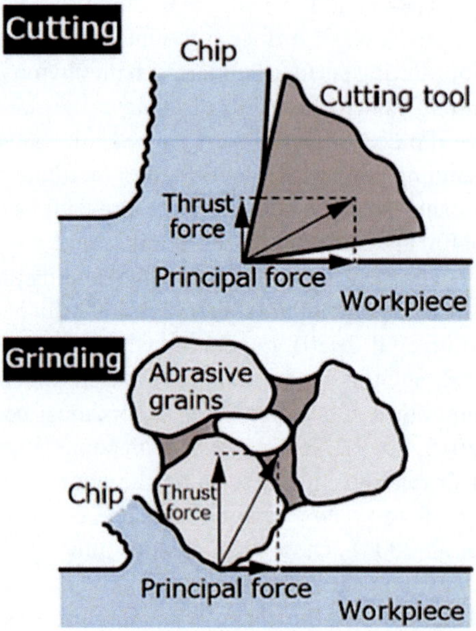

7.3 Sanding Machine Technology

The abrasive sanding process can be carried out manually, semi-automatically, and/or fully automatically. The automatic sanding machines usually have two or three sanding heads (usually attached with sanding belts), in which the first sanding belt is rougher than the second and third. The machine comes attached with a dust exhaust system to capture and remove dust particles, as air pollution control measure (Csanády et al. 2019).

In the abrasive sanding process, the important process variables taken into consideration are the stock removal rate, feed rate, platen pressure, nature of sandpaper used,

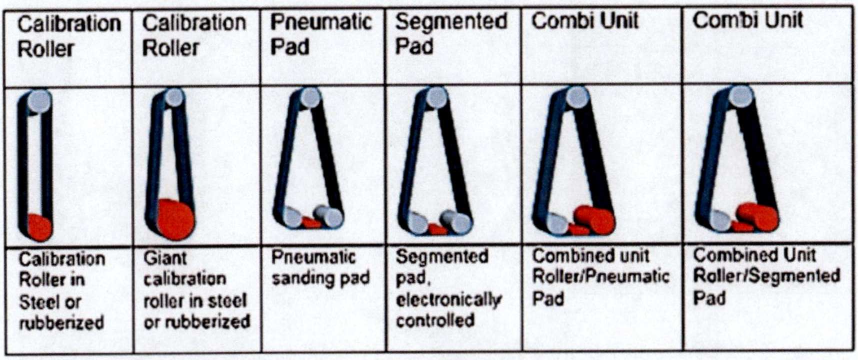

Fig. 7.6 Design of pressure mechanism in sanding machines (Courtesy of the Malaysian Furniture Council)

and the workpiece characteristics. The application of a moderate cutting speed (10–15 m/s) with relatively higher platen pressure may guarantee a much better utilization of the abrasive belts. Platen pressures between 0.8 and 1.0 N/m^2 are considered as optimum pressure value (Ratnasingam et al. 2004). The different sanding head combinations found in the different type of wide-belt sanding machine are shown in Fig. 7.6, and each is meant for different application. Calibration sanding machines have heavy-duty applications and are usually found in wood-based panels (i.e., particleboard and medium density fiberboard) factories or edge-glued laminated board-producing factories (Ratnasingam et al. 2005).

The semi-automatic sanding machines usually have one operator controlling the sanding process manually but is also equipped with a dust exhaust system. The abrasive sanding elements used in such machines may be brushes, sheets, or pads, or a combination of all (Ratnasingam et al. 2004; Csanády et al. 2019). Nevertheless, these machines must be used more cautiously to avoid inflicting further blemishes and sanding marks on the surfaces of the workpieces (Fig. 7.7).

From a theoretical perspective, abrasive sanding being an orthogonal cutting process (i.e., in which the cutting edge is perpendicular to the relative motion of the tool and workpiece) generates a surface which may be a plane parallel to the initial work surface. The oscillation of the abrasive sanding belt in sanding machines, introduces a slight change to the orthogonal cutting principle (Ratnasingam et al. 2004). This reverse oscillation is important to avoid the roll off of the sanding belt from the sanding heads (Fig. 7.8). Moreover, the additional cross-sanding unit, which results in an oblique-sanding mechanism, produces uniform surface gloss, while minimizing visible surface scratches on the workpiece.

Nevertheless, the outcomes of the abrasive sanding process in furniture manufacturing are subjective and dependent on the operator and the process parameters used. The results may be variable depending on the experience of the operator but may be optimized by following some basic guidelines (Ratnasingam et al. 1999). To achieve the best results in furniture and wood products manufacturing, sanding is usually

Fig. 7.7 Common sanding machines (Courtesy of the Malaysian Furniture Council)

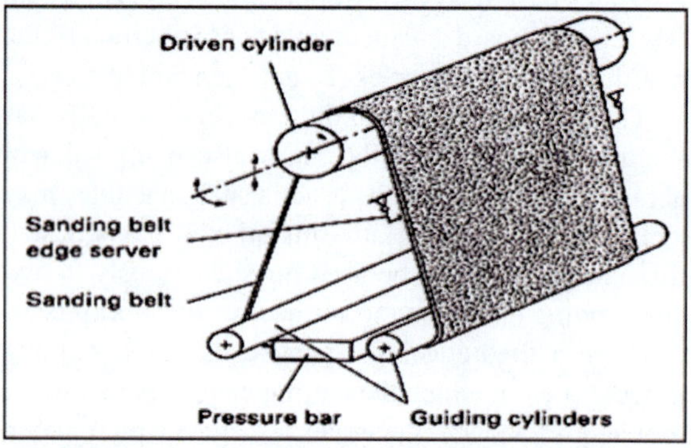

Fig. 7.8 Sand paper stabilization mechanism in wide-belt sander

started off with coarse abrasive grit sizes, and gradually changing to medium grit sizes, and then to fine grit sizes, as the work progresses (Luo et al. 2020). Although it is not necessary to use every available grit size, it is important that the difference between successive grit sizes used is not too large (i.e., large sequential grit sizes), to avoid improper sanding outcomes. This is because skipping too many grit sizes may leave scratches that cannot be removed by the subsequent grit size, and often requires more time to smoothen it out (Nagyszalanczy 1997; Gurau et al. 2005). In essence, the correct coated abrasive and optimum set of process parameters are a necessity for a productive abrasive sanding operation. The economics of the abrasive sanding process is as illustrated in Fig. 7.9, where optimum cost obtained between

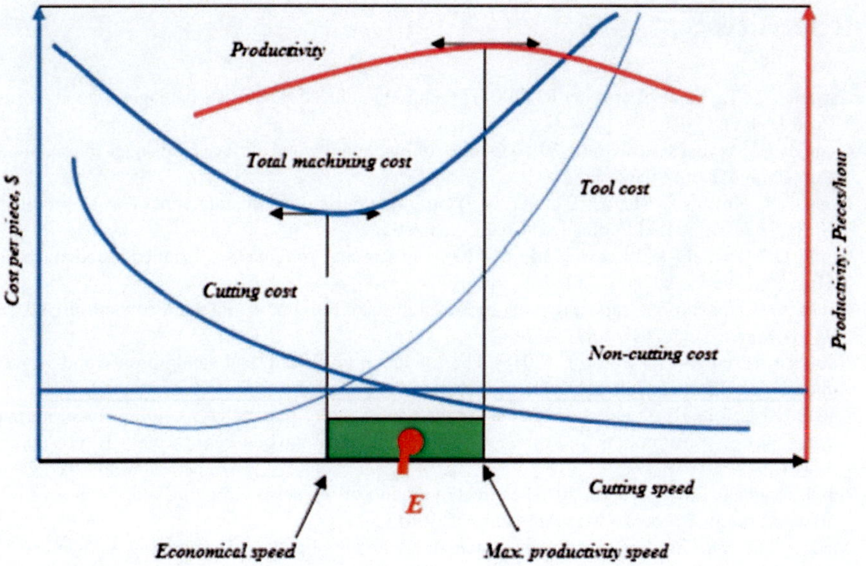

Fig. 7.9 Productivity in abrasive sanding process

moderate cutting speed and throughput rate is used (Taylor et al. 1999; Miao and Li, 2014).

In essence, the abrasive sanding process is often considered part of the overall 'cosmetic operation' in furniture manufacturing, as it prepares the furniture for its aesthetic appeal and quality finish. Nevertheless, blemishes and defects in abrasive sanding may also arise, and all efforts must be taken to ensure that these quality impairments are minimized. A detailed discussion of the quality problems encountered in the abrasive sanding process is covered in Ratnasingam et al. (2004, 2005), and the readers are advised to be familiar with the manifestation of these defects and blemishes, as well as the remedial measures to overcome it, in order to optimize the productivity of the abrasive sanding processes.

Summary

- The sanding process is the foundation for a good finish or coating application.
- It aims to flatten and smoothen the surface, removing blemishes from the previous operations.
- It has a significant effect on the aesthetic appeal of the furniture, as it is the pre-requisite for a good cosmetic operation.

References

Carrano AL, Taylor JB, Lemaster RI (2004) Machining induced subsurface damage to wood. Forest Prod J 54(1):85–91

Csanády E, Magoss E, Tolvaj L (2015) Quality of machined wood surfaces. Springer International Publishing Cham, Switzerland

Csanády E, Kovács Z, Magoss E et al (2019) Optimum design and manufacture of wood products. Springer International Publishing, Cham, Switzerland

Gurau L, Mansfield-Williams H, Irle M (2005) Processing roughness of sanded wood surfaces. Holz Roh Werkst 63(1):43–52

Hendarto B, Shayan E, Ozarska B et al (2006) Analysis of roughness of a sanded wood surface. Int J Adv Manuf Tech 28(7–8):775–780

Kúdela J, Mrenica L, Javorek Ľ (2018) The influence of milling and sanding on wood surface morphology. Acta Facultatis Xylologiae Zvolen 60(1):71–83

Luo B, Li L, Liu H, Xu M et al (2014) Analysis of sanding parameters, sanding force, normal force, power consumption, and surface roughness in sanding wood-based panels. BioResources 9(4):7494–7503

Luo B, Zhang J, Bao X et al (2020) The effect of granularity on surface roughness and contact angle in wood sanding process. Measurement 165:108133

Malkin S (1989) Grinding technology—technology and application of machining with abrasives. Ellis Horwood Publication, London, United Kingdom

Miao T, Li L (2014) Study on influencing factors of sanding efficiency of abrasive belts in wood materials sanding. Wood Res 59(5):835–842

Murmanis L, River BH, Stewart HA (1986) Surface and subsurface characteristics related to abrasive-planing conditions. Wood Fiber Sci 14(3):106–109

Nagyszalanczy S (1997) The wood sanding book—a guide to abrasives, machines, and methods. The Taunton Press, Newtown, Cincinnati, USA

Očkajová A, Kučerka M, Krišťák L et al (2018) Granulometric analysis of sanding dust from selected wood species. BioResources 13(4):7481–7495

Porankiewicz B, Banski A, Wieloch G (2010) Specific resistance and specific intensity of belt sanding of wood. BioResources 5(3):1626–1660

Ratnasingam J, Reid HF, Perkins MC (1999) The productivity imperatives of coated abrasives application in furniture manufacturing. Holz Roh Werkst 57:117–120

Ratnasingam J, Reid HF, Perkins MC (2002) The abrasive sanding process of Rubberwood (Hevea brasiliensis): An industrial perspective. Holz Roh Werkst 60:191–196

Ratnasingam J, Scholz F (2004) Optimizing the abrasive sanding process of Rubberwood. (Hevea brasiliensis). Holz Roh Werkst 62(6):411–418

Ratnasingam J, Scholz F, Friedl E (2004) Wood sanding processes—an optimization perspective. UPM Press, Serdang, Malaysia

Ratnasingam J, Scholz F, Wong D et al (2005) The sanding of wood-based panels. UPM Press, Serdang, Malaysia

Ratnasingam J, Scholz F (2008) Reducing fuzziness in abrasive sanding of Rubberwood (Hevea brasiliensis). Holz Roh Werkst 66(2):159–160

Ratnasingam J, Scholz F, Natthondan V (2011) Dust generating characteristics of hardwoods during sanding processes. Eur J Wood Wood Prod 69(1):127–131

Ratnasingam J, Lim CL, Hazirah AL (2019) A Comparison of the abrasive sanding dust emission of oil palm wood and rubberwood. BioResources 14(1):1708–1717

Sofuoğlu SD, Kurtoğlu A (2015) Effects of machining conditions on surface roughness in planing and sanding of solid wood. Drvna Ind 66(4):265–272

Taylor JB, Carrano AL, Lemaster RL (1999) Quantification of process parameters in a wood sanding operation. Forest Prod J 49(5):41

Chapter 8
Furniture Finishing

The finishing operation in furniture manufacturing is the most important value-adding operation, which not only imparts the aesthetic appeal, but also protection to furniture. It is a regarded a cosmetic operation, which reveals the beauty of furniture, enticing customers to purchase it. Finishing is done using one of the many different types of chemical coatings available in the market, applied carefully though the appropriate process, and cured sufficiently. The development in coating chemistry and application technology has transformed the finishing operation. With growing concern for environmental compliance, the traditional solvent-based coatings are being replaced by water-based coatings throughout the world. This chapter discusses the nature of the finishing operation, the materials used, and the application techniques.

8.1 Introduction

The finishing operation in furniture manufacturing is considered the most important value-added operation, which predetermines the aesthetic appeal and marketability of furniture. It is considered a cosmetic process, in which the finish or coating material is applied on to the piece of furniture to enhance its aesthetic appeal, protecting it and enhancing its value to attract potential customers (Fig. 8.1). Studies have shown that the finishing process may represent between 5 and 30% of the total furniture manufacturing costs (Tichy 1997).

Finishing is the final process in furniture manufacturing and is the most important value-adding operation, which imparts the furniture with desirable characteristics, including enhanced aesthetic appeal, increased surface hardness, as well as improved resistance to environmental agents, household chemicals, and liquids (Flexner 2010). Finishing can also make the furniture easier to be cleaned, while keeping it sanitized as well as sealing the pores to minimize breeding grounds for micro-organisms. Further, finishing provides value addition, in which it could change the appearance of low-value woods, making it look more expensive, and of higher value (Tichy

J. Ratnasingam, *Furniture Manufacturing*, Design Science and Innovation,
https://doi.org/10.1007/978-981-16-9412-7_8

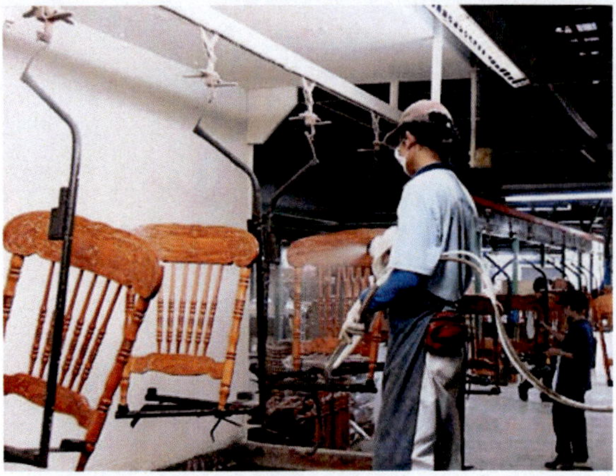

Fig. 8.1 Furniture finishing operation (Courtesy of the Malaysian Furniture Council)

1997). For detailed discussion on wood finishing/coating processes, the readers are referred to Bentley and Turner (1998), Bulian and Graystone (2009), Jones et al. (2017), and Prieto and Kiene (2018), which provides a thorough insight into the chemistry of coatings, its application, and performance. Being organic coatings, wood finishing/coatings are subjected to formation of defects, both during application and drying/curing. In this respect, the publication by Hamburg and Morgans (1979) provides the most comprehensive discussion on paint film defects, its causes and cure.

The furniture finishing process involves many steps, and a perfectly finished furniture is usually the outcome of a tedious process of preparing the furniture surface (i.e., sanding process), then using suitable finish or coating material, which are applied through the correct techniques and allowed to cure or harden thoroughly. In fact, to produce a perfectly finished piece of furniture it takes patience and ample care (Tichy 1997).

Technically, a good furniture finishing operation involves both mechanical and chemical processes (Csanády et al. 2019). The mechanical processes refer to several preparation steps, including sanding, filling up with putty/filler, wiping of stain with a rag, and applying several layers of finishes either by spraying or simply brushing it on. On the other hand, the finish materials used are considered as chemical processes, which determines the choice of finish or coating materials to be applied, including, dyes, stains, oils, varnishes, lacquers, solvents, etc. Inevitably, every finish has its own distinct properties, which confers characteristic surface finish quality on the furniture (Kieblesz et al. 2017).

8.2 Finishing Steps

The process of furniture finishing starts with preparing the surface for the finish or coating application. The quality of the final finish will be impaired if the surface preparation is poor, and it is contaminated with dust, particulates, etc. Therefore, a clean, flat, and smooth surface is a necessity for the subsequent finish or coating materials application (Suleman and Rashid 2007; Slabejová et al. 2016).

The preparation of the furniture surface usually involves some sanding (usually using grit sizes of 240 or 280), aimed at removing defects, flattening, and smoothening the surface. In many instances, the surface is thoroughly cleaned, usually with a good wipe down or using air blowers to clean the surface. This step is usually carried out away from the finishing section, which is often pressurized to maintain a clean, dust-free environment (Kúdela and Liptáková 2006).

After this step, undesirable defects on the surface, such as gaps from knots or nails, are concealed by filling these crevices, and gaps with wood putty, wood fillers, and, in cases of bigger gaps, a combination of both. In some instances, fine wood dust mixed with cyanoacrylate glue/adhesive is used to fill up deep cracks and dents on the surface, which conceals the blemishes well.

Next is the bleaching step, if necessary. Sometimes, the color from stained piece of wood or the natural color in a piece of wood will have to be removed. In either case, a wood bleach made up of sodium hydroxide with hydrogen peroxide is usually used. Other common wood bleaches available in the market are chlorine bleach and oxalic acid, but their effectiveness is somewhat limited (Flexner 2010). Bleaching of wood surfaces is required, when a new color needs to be imparted on the surface or a different color tone required.

This is followed by the coloring step, which is achieved through the application of stains (i.e., penetrating-type dye stains or suspension-type pigment stain) and colored wax, by wiping or brushing the stain on the surface (Hse and Kuo 1988). The wiping must be done evenly to avoid blotchiness. Once the staining is balanced out, and the desired color achieved, the workpieces are left to dry out for the next steps.

Depending on the desired finish/coating as specified by the customer, a suitable choice of finish or coating is applied usually in several layers/coat on the furniture. Light inter-coat abrasive sanding is carried out to enhance the anchorage of the next coat of finish/coating material. The number of layers or coats applied will depend on the type of finish desired. For instance, in the typical American finish, up to 15 steps are involved before the final finish quality is achieved. This is obviously time-consuming as each layer must be dried, cured, and sanded before the next layer is applied. Finally, the final top-coat finish or coating layer is polished, or buffed using steel wool, or pumice to produce a shiny or glossy surface (Slabejová et al. 2016).

The shine or gloss of the finished surface of furniture is a desirable aesthetical feature sought after by most customers. Shine or gloss can be imparted through different methods. The variables affecting the gloss level are material density, machining process, thickness, and flatness of the finish/coating layer applied (Csanády et al. 2019). Research has shown that conventional knife machining gives higher natural gloss level, but the variations in gloss level could be attributed to the

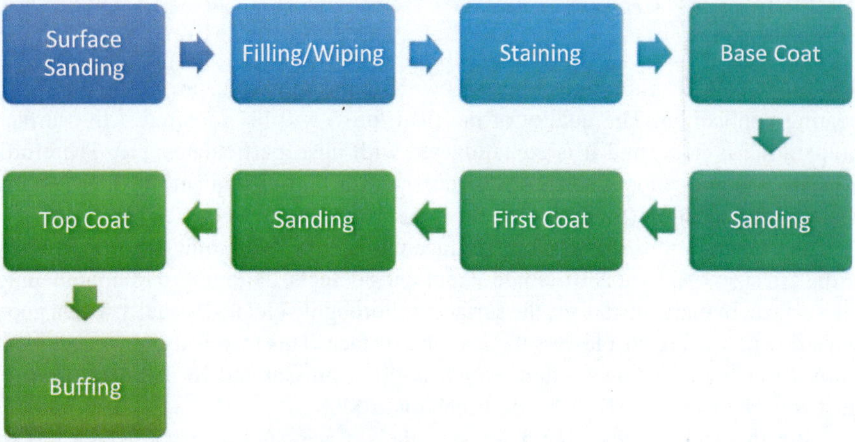

Fig. 8.2 Common furniture finishing steps

anatomical differences in the wood (Slabejová et al. 2016). On the other hand, abrasive sanding, especially perpendicular to the grains, often results in more uniform gloss levels, with low gloss value (Ratnasingam and Ioras 2013). However, the natural gloss of wood surfaces varies with time and may not be preserved throughout the service life of the furniture. Therefore, in most furniture factories, finish or coating materials with the required gloss level with minimum variations are commonly used to meet customer expectations (Ugulina and Hernãndez 2018).

For color enhancement of furniture, the surfaces may be treated with oils, lacquers, wax, or a blend with shellac. The color hue remains much the same, but its lightness diminishes, while the color saturation increases significantly. The sanding process imparts a gray shade over the wood surface due to the broken cell walls, which diffuses the reflection of light (Csanády et al. 2019). However, it must be emphasized that the abrasive sanding process usually produces a matt surface which, when applied with lacquer of a given gloss level, produces an even surface glossiness, which is highly desirable in furniture. Ensuring uniform glossiness is important especially in high-end furniture, as it is an aesthetic appeal demanded by customers (Ratnasingam and Scholz 2006).

In most furniture factories, a process chart depicting the planned finishing steps in the specific sequences will be drawn and shown to all operators, to ensure consistency in the final finish quality (Fig. 8.2). The fact that different type of finish/coating material imparts different levels of durability and protection to the furniture, other factors such as the ease of application, performance, and cost are also taken into consideration when finding the most suitable finish/coating material for the intended purpose.

Further, the finishing sequence also has a strong influence on the manufacturing lead time and the cost involved, which, in most instances, are high on the priority list of furniture manufacturers.

8.3 Properties of Finish Materials

The binder or resin used in the formulation usually lends its name to the finish/coating, such as urea, epoxies, alkyds, urethanes, etc. It must be emphasized that resins are not the only constituent of a coating. The coating formulation consists of five important ingredients, i.e., resin, additive, extender, pigment, and solvent (Bentley and Turner 1998). Additives lend the coating certain performance properties, while extenders serve as bulking agents or to make up the solids in the coating film. The pigments are the coloring agent in the coating. All these ingredients are dissolved and mixed into a liquid for easy application, and the mixing agent is often a mix of chemical solvents or water. There are many types of finish or coating materials available in the market for furniture application (Table 8.1), but a few types are more extensively used in the furniture industry due to its distinct advantages and market preferences (Flexner 2010).

With growing concern for environmental protection and the need to reduce emission of chemical solvents, **water-based coatings** have been developed by the global coating industry. As the name implies, a proportion of water is used as the solvent in water-based coatings, but is not free of solvents completely, though. Many water-based coatings contain co-solvents, which are essentially low concentration solvents, which assist to push the water out during the drying of the coating. Since water-based coatings are either free or with very little solvents, it has emerged as an important alternative offered by the coatings industry to reduce the volatile organic compounds (VOC) emission, both from the resin and the solvents (Tichenor and Guo 1991). Traditional solvent-based coatings dry through the evaporation of the solvents, often initiated by a chemical reaction with oxygen in the atmosphere. Typically, moving air will speed up the drying of solvent-based coatings, thereby reducing drying times. Inevitably, solvent-based coatings have one major advantage over water-based coatings. They are less vulnerable to atmospheric conditions, such as temperature and humidity, when curing. Variable environmental humidity can slow the drying of water-based coating significantly, making it less useful in some climates, apart from the problems related to grain-raising in wood materials, due to the hygroscopic nature of wood (Flexner 2010).

Table 8.1 Properties of common furniture finishes

Type of finish	Appearance	Protection	Ease of application
Wax	Dull with even sheen	Short term	Easily applied with cloth or brush. After curing it is buffed with a cloth to the desired sheen level
Shellac	From clear to rich orange color	Average water resistance, but poor resistance against solvents	Ox or badger/skunk hairbrush or pad is recommended for application. Difficult to be sprayed or brushed
Linseed oil	Yellow warm glow and darkens with age	Low	Easy applied by simply brushing or wiping. Takes long time to cure
Alkyd varnish	Yellowish orange tint	Reasonably good protection	Moderately easy to apply, either by brushing, roller coating, or spraying
Nitro-cellulose lacquer (NC)	Transparent, satin and gloss	Reasonable protection	Moderately easy to apply, and usually sprayed if large volume is involved. No inter-coat sanding is required
Pre-Cat lacquer	Transparent, all sheens from 5 to 90%	Reasonably good resistance against wet and dry heat	Moderate. Requires spray equipment. No sanding required between coats
Acid-cured lacquer	Transparent, all sheens from 5% to gloss	Excellent resistance against most chemicals	Moderately easy to apply, but requires spraying, with inter-coat sanding
Polyurethane water-based varnish	Transparent or colored, with sheens up to 80%	Excellent durability and UV stable	Easy to apply by spraying and requires inter-coat sanding. Cures relatively fast
Two-part polyurethane lacquer	Transparent	Stronger protection than regular polyurethane varnish	Usually sprayed, and equipment must be cleaned immediately
Epoxy coating	Thick, high-gloss, and transparent. However, clouding or yellowing upon UV exposure possible	High level of protection	Easy pour-on application for flat surfaces, but difficult to apply evenly on complex profiles/shapes

Adapted from Bentley and Turner (1998)

8.4 Powder Coating

Powder coating is revolutionizing the wood finishes or coatings market, especially for medium density fiberboard (MDF). Powder coating technology offers opportunities that other finishing and laminating systems do not offer, which is highly desirable when furniture is regarded fashion (Tichy 1997). Powder coating gives an aesthetically appealing, strong, and seamless finish in colors of the rainbow. Further, MDF is protected from blemishes which may be caused by chipping, stains, spills, and scratches, by the strong and durable powder coating. As a matter of fact, MDF is appropriate for powder coating application due of its low porosity, homogeneous surface, and its sufficient and uniform moisture content which facilitates conductivity, thus allowing it to be coated directly. To improve electrostatic conductivity during application, MDF is sprayed with a solution that gives a better conductive surface. The MDF is then pre-heated to the desired temperature that helps melt the powder when it is applied, thus adhering to the surface. A uniform surface temperature on the MDF is important to ensure a high transfer efficiency during application, and a uniform appearance. In commercial powder coating applications, the charged powder is deposited effectively on to the surface of the MDF, as the spray gun is connected to an electrostatic charge (Jocham et al. 2011). Powder materials for MDF are thermally cured, usually made up of ultra-violet (UV) cured powders. Thermally cured powders require either infrared ovens, convection ovens, or hybrid ovens that combine both infrared and convection heating. The thermal energy dissolves the powder so it will flow to form a uniform and consistent finish film, through curing or crosslinking.

8.5 Curing of Finishes

Finishes and coatings used for furniture are also categorized by the method they cure or harden (Table 8.2). There are three different mechanisms by which furniture finishes/coatings cure:

Table 8.2 Curing methods for common wood finishes

Finish	Curing
Shellac	Evaporative
Lacquer	Evaporative
Wax	Evaporative
Varnish	Reactive
Oil	Reactive
Powder coat	Reactive
Water based	Coalesce

Adapted from Csanády et al. (2019)

(1) Evaporative: These types of finishes/coatings cure when the solvent evaporates, leaving a dry resinous film on the surface.
(2) Reactive: These types of finishes/coatings dry through a chemical modification process, also known as polymerization, which results in a dry film that is not easily dissolved by solvents.
(3) Coalesce: These types of finishes/coatings dry by a combination of evaporative and reactive processes. A good example is the water-based finishes. They are essentially emulsions with slow-evaporating solvents, such as glycol-ether and water (Csanády et al. 2019).

Generally, finishes/coating dry and cure faster when the surrounding temperature is increased. Hence, hot-air ovens in the form of a room, or a tunnel (with gradually increasing temperatures of up to 60 °C, with a stream of air at 0.5 m/s) are usually used in the furniture finishing section. However, conventional hot-air ovens are not suitable for water-based finishes/coatings, as water takes much longer to evaporate than solvents (Fig. 8.3). For example, if the drying length must remain the same because of space limitations, then to cure water-based coatings using the existing hot-air oven, the flow speed through the oven must be significantly reduced to increase dwell time. This can lead to production bottlenecks, which may not be desirable in the furniture factory (Tank 1991).

In high-volume furniture production factories, conveyorized finishing lines are commonly used to allow a continuous production system with variable speed (Bentley and Turner 1998). The overhead and palletized floor lines are the two common types of conveyorized finishing lines used in furniture factories (Fig. 8.4).

The design of the conveyorized finishing line must consider the typical finishing steps taken, floor space available, number of spraying stations, and the range of speed desired (Tank 1991). Further, it is also necessary to induce a positive pressure in the finishing section, to ensure a dust-free environment, and to achieve the highest quality finish.

One of the common problems in wood finishing/coating in furniture manufacturing is insufficient heating (Tsotsas and Mujumdar 2011). A practical solution

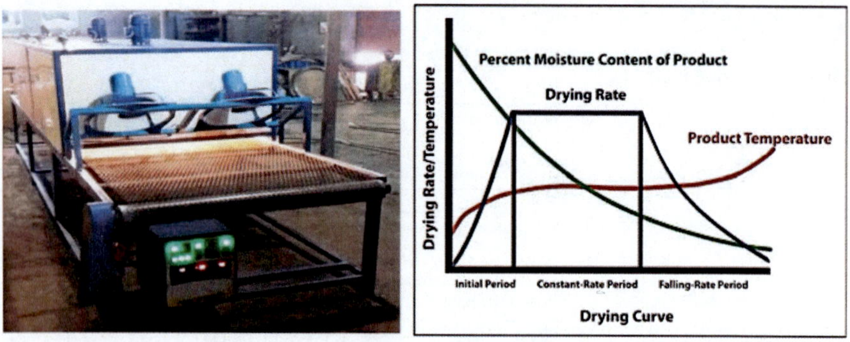

Fig. 8.3 Drying oven and the drying curve (Courtesy of the Malaysian Furniture Council)

Fig. 8.4 Conveyorized finishing lines (Courtesy of the Malaysian Furniture Council)

to this problem is offered by infrared technology. Infrared emitters and systems, because of their compact design, can easily be retrofitted within existing hot-air drying systems. The wavelength of infrared radiation has a strong influence on the curing of water-based coatings. Water evaporates rapidly when heated with medium-wave infrared, as it is converted more effectively into heat energy. Because of this characteristic, carbon emitters have been developed specifically for medium wave. Infrared emitters incorporating carbon technology can provide power densities up to 150 kW/m^2 and short response times. Carbon infrared emitters combine medium-wave radiation with high power densities to accelerate the drying of water-based paints and lacquers in the furniture industry (Prieto and Kiene, 2018).

Infrared heating technology offers several solutions for energy optimization in industrial heating processes, including the following:

- High heat exchange capacity.
- Contact-free heat transfer.
- Efficient energy exchange through ideal optimum wavelengths.
- Use of energy only when required due to fast response times.

Therefore, modern infrared heaters are increasingly being used in furniture manufacturing factories to increase throughput. However, the selection of the type of infrared heaters must be matched with the products and processes, not only in terms of wavelength, but also its control and emitter shape (Tsotsas and Mujumdar 2011). Nevertheless, for every furniture factory, it is worthwhile to select the heating source to be used, based on the characteristics of the processes and materials used. This not only ensures a higher throughput rate on the factory shopfloor, but also improves product quality and lower manufacturing cost.

8.6 Finish Application Methods

Finishes or coatings for furniture can be applied through several methods, including dipping, roller coating, and spraying.

Dipping can be carried out by two different methods. The first method involves dipping of the whole piece of furniture into the tank filled with the finish. By doing so, the finish will coat all the surfaces, including gaps and cuts, although an even application may not be achieved. In this process, the excess of the drip off is collected and reused, is channelled into a diverse tank, and is reused. However, the most important drawback is that the tank must be filled with a large volume of finish, which must be maintained throughout the process. This disadvantage can be reduced significantly by pouring the finish/coating on to the object, but it is time-consuming (Talbert 2008). Second is the controlled dipping method, where the furniture is dipped into the tank and withdrawn to allow excess finish/coating to drain off. If carried out carefully, this method avoids finishing/coating defects, such as lacquer-run or sagging (Fig. 8.5). The viscosity of the finish/coating is important in the dipping method, as the flow down of the excess finish/coating is inversely dependent on its viscosity, while the thickness of the finish/coating remaining on the surface is proportional to the initial shear stress τ_0 required to start the flow (Csanády et al. 2019).

Roller coating is a process in which rotating rolls in contact with the substrate apply the finish/coating onto the surface. Roller coating is designed for flat and level surfaces and is not appropriate for three-dimensional components, shaped, or profiled panels (Csanády et al. 2019). The main advantages of roller coating are they are fast and efficient, making this technology an excellent choice for mass production of flat-panel furniture. This method ensures a consistent and uniform quality application, when the roll gaps are controlled tightly, thus giving high productivity especially when the system is completely automated. These machines usually come with two-head configuration, an applicator roll (*2*), which applies the coating onto the substrate and a dosing roll (*1*), which controls the amount of finish/coating to be applied (Fig. 8.6).

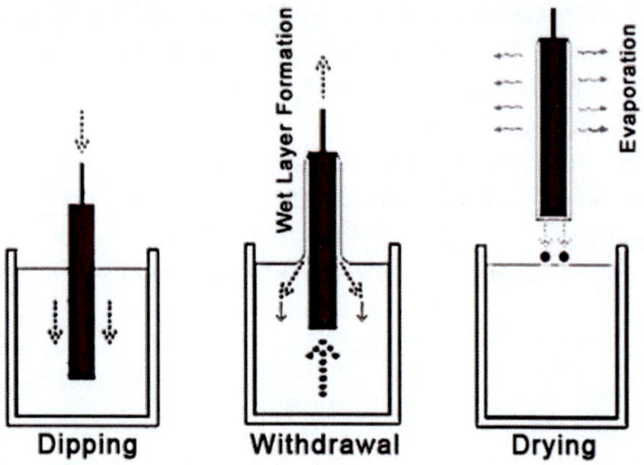

Fig. 8.5 Dipping technology

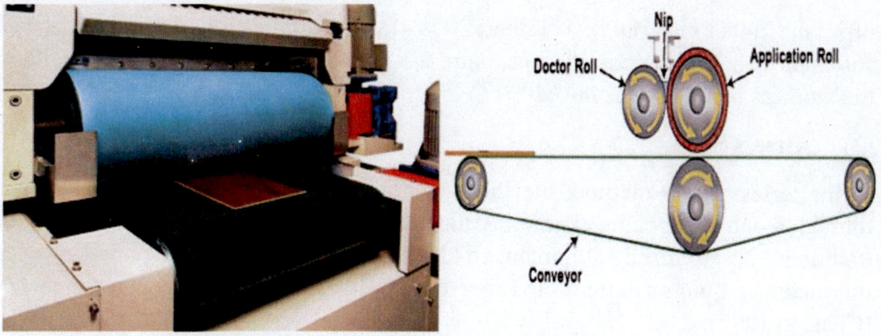

Fig. 8.6 Roller coating technology

Spraying is the most prevalent coatings application technique in the furniture manufacturing industry, due to its flexibility and financial viability (Talbert 2008). There are a few spray methods available in the market which are given below:

- Air atomized.
- Airless.
- Air-assisted airless.
- Electrostatic.
- High volume, low pressure.

(1) **Air-Atomized Spray**

Air-atomized application is the traditional method of spray technology. It uses a conventional spray gun that combines compressed air with a fluid stream, to form a pressurized fog/mist that coats the furniture (Charron 1996). This method is highly flexible and suits many different applications and substrates, as the equipment can be adjusted and customized to the needs of the operator (Fig. 8.7).

However, the downside of the air-atomization spray method is loss of coating. This equipment has low transfer efficiency, although the savings in labor cost may

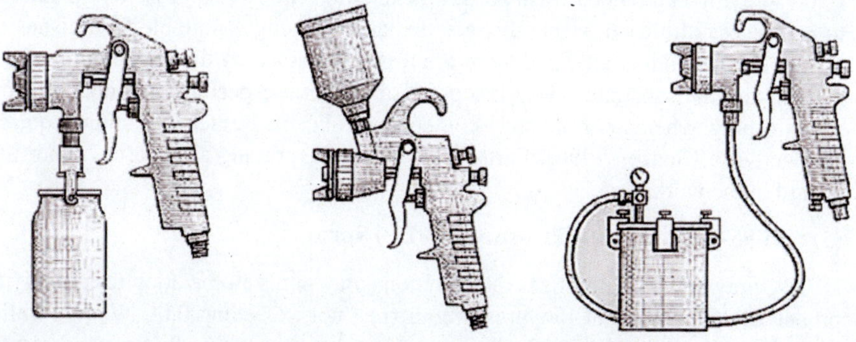

Fig. 8.7 Spray gun technology

offset the higher material cost (Talbert 2008). In such applications, spray booths with sufficient exhaust are required to capture the excess coating to ensure the health of the workers is not compromised.

(2) **Airless Spray**

In the airless spray method, the finish/coating fluid is pressurized by pushing it through a narrow opening. The spray mist travels at a lower speed compared to the traditional air-atomized equipment, so there is less loss due to overspray. However, this method requires a more skilled operator as well as higher equipment maintenance (Charron 1996).

(3) **Air-Assisted Airless Spray**

This spray method is most widely utilized within the furniture manufacturing industry. It employs a high-pressure liquid supply for atomization and compressed air at the cap for spray pattern control. Air-assisted airless spraying solves many of problems faced when applying high-viscosity and high-solid finish/coatings, as it offers the advantages of both the air and airless spray systems (Charron 1996). Nevertheless, it requires higher maintenance and operational costs.

(4) **Electrostatic Spray**

Electrostatic spray application is the most advanced spraying method, with very high transfer efficiency. This method is based on the principle that opposite electrical charges attract each other. The coating is electrostatically charged, because it passes through an electrostatic field created between the anode at the front of the gun and a grounded object, which is the workpiece. The charged coating particles are attracted to the grounded workpiece, forming a layer of coating. The charged material will wrap itself evenly around the workpiece, which accounts for the higher transfer efficiency (Bulian and Graystone 2009), resulting in a smooth, uniform, and solid layer of coating, even on profiles and sharp corners (Fig. 8.8).

In high-tech, large volume furniture factories, the rotary atomization electrostatic spraying or centrifugal spraying, which is another form of air-atomization spray is used to meet the high throughput rates required (Jones et al. 2017). While electrostatic spraying can reduce wastage significantly, it requires expertise and precision to be successfully used. Generally, electrostatic spraying is suitable for finishes or coatings with conductivity, and therefore it is constrained by the formulation of the coating and the substrates. This is a point of concern, especially in wooden furniture finishing, where poor moisture content control can adversely affect the required conductivity (Charron 1996). Further, electrostatic spraying also poses higher fire hazard in the workplace.

(5) **High-Volume, Low-Pressure (HVLP) Spray**

HVLP spray equipment utilizes the common spray gun's atomization technology to propel a large amount of low-pressure air (i.e., not exceeding 0.06 N/mm^2) to the spray gun. Since the coating is discharged at a lower velocity, this method results in less overspray and blowback than the traditional air-atomized method (Charron

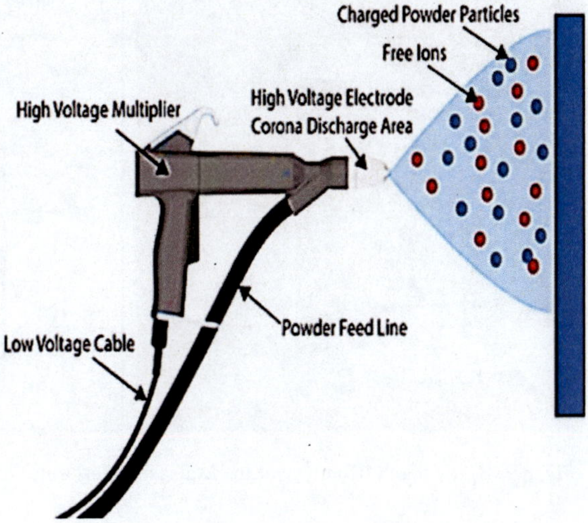

Fig. 8.8 Electrostatic spraying

1996). While the method is advantageous, the HVLP method required highly skilled operators to ensure its high transfer efficiency.

In contrast the **Low-Volume, Medium-Pressure** (LVMP) spray method uses air pressures not exceeding 0.19 N/mm^2 at the air inlet (Talbert 2008). This permits the use of an air cap, designed to give high-quality finish, while providing transfer efficiency higher than the HVLP method (Table 8.3).

One of the most important maintenance and pollution control equipment in the finishing section in furniture factories is the **spray booth**. A spray booth acts as a pressure-controlled closed environment used in finishing sections to minimize pollution from overspray and reduce VOC emission (Tank 1991). To ensure optimal working conditions (temperature, airflow, and humidity), spray booths are equipped with ventilation, comprising of mechanical fans driven by electric engines, and alternatively with burners to heat the air, to hasten the drying of the coating (Fig. 8.9). As a guideline, airflow speed across the surface of the spray booth should be maintained 0.5 m/s, while the air exchange volume at the spray booth, with an area 7.5m^2, should be approximately 3.5m^3/s (Talbert 2008). This is to ensure toxic solvents

Table 8.3 Comparison of different spraying systems

Method of application	Capital cost	Process complexity	Waste and emissions	Additional considerations
HVLP spray	Low	Low	Medium	–
Air-assisted airless spray	Low	Low	High	–
Airless spray	Medium	Low	High	–
Electrostatic spray	Medium	Medium	Medium	Only conductive parts can be painted

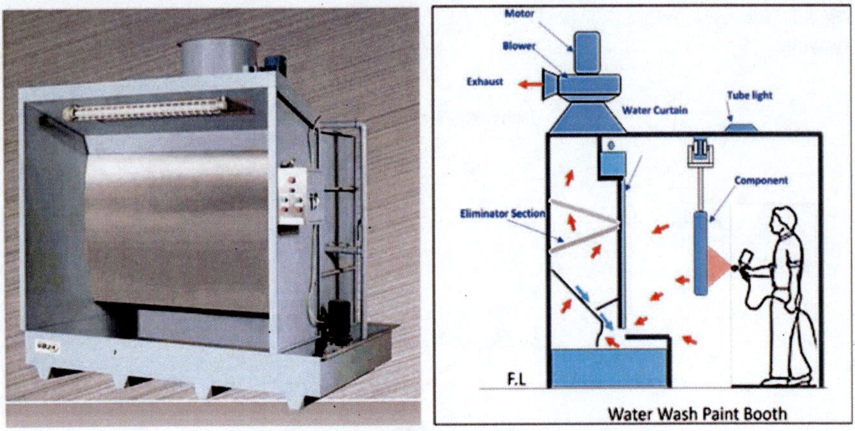

Fig. 8.9 Spray booth (Courtesy of the Malaysian Furniture Council)

and coating particles are exhausted to the outside environment, after filtration, and necessary treatment to reduce air pollution. Fires and explosions due to dust is also prevented by the spray booths. Water-washed spray booths utilize specialty chemicals to minimize tacking of the coatings, and a down-draft air to effectively capture overspray, while reducing air pollution (Tichy 1997).

8.7 Transfer Efficiency and Finish Application

The economics of finishing/coating operation in furniture manufacturing is dependent on the effectiveness at which the finish/coating is applied, which refers to the **transfer efficiency.** Improvements in transfer efficiency during applications results are reduced wastage of finishes/coatings as well as lowering emission of volatile organic compounds (VOCs) (Talbert 2008). In practice, transfer efficiency is influenced by several parameters, in which some are within the control of the operator, whereas others are not. The key parameters include the following:

- The choice of spray system used.
- Target setup, configuration, and size—higher transfer efficiency is easier to achieve on flat surfaces, as opposed to shaped or profiled products.
- Spray booth setup—stray cross-drafts and downward-drafts tend to increase wastage, by diverting the spray mist away from the target object. Further, transfer efficiency of the electrostatic spray system is also influenced by temperature and humidity control in the finishing facility.
- The coating characteristics.

- Coating and air flow rates—spray guns should be operated under optimum flow rates. Surpassing the recommended flow rates often leads to lower transfer efficiency, due to increased incidence of blowback (i.e., coatings bouncing off the workpiece) and overshoot. Excessive air pressure may cause premature drying of the coating, even before it reaches the surface of the object (i.e., coating fog) (Charron 1996).
- Spray gun distance from the workpiece—when the distance between the spray gun and the workpiece is too close, coating bounce-back increases, resulting in lower quality finish. Finish defects, such as sags and runs, are common under this circumstance. On the other hand, if the distance between the spray gun and the workpiece is too far, overshooting and finish fogging are commonly encountered.
- Operator error.

In a technical perspective, transfer efficiency is defined as the volume of finishes/coatings solids deposited on the workpiece, divided by the volume of finishes or coatings solids sprayed at the workpiece, multiplied by 100%. However, this concept of transfer efficiency excludes several related factors that influence coating material utilization. Nevertheless, it must be kept in mind that wastage cannot be reduced solely by using application techniques with the highest rated transfer efficiencies (Talbert 2008).

However, the actual transfer efficiency during spray application is dependent on several factors as described by Charron (1996):

(1) **Quality of Finish.** As a rule, the quality of the finish improves, as the size of the spray particles reduces. Unfortunately, as the size of spray particles decreases, transfer efficiency also decreases. Some of the finest particle sizes are achieved with low-volume high-pressure (LVHP) spray system; however, this is the least efficient means of applying coating. To meet the finish requirements, a compromise must be reached between transfer efficiency and quality of finish.

(2) **Production Rate.** Production managers and engineers must agree to the desired production rate on the factory shopfloor, before deciding the transfer efficiency of the coating system, especially if coating is being done on a conveyorized system that includes other operations. This is because the efficiency of spray devices will vary with the rate of application.

(3) **Desired Coating Film Thickness.** To determine the optimum transfer efficiency, the thickness of the applied coating film versus the thickness of the cured coating film should be taken into consideration. For instance, if a 1-mil (or 25 μm)– thick dry coating film is required, but the spray technique can deliver coating film of 2 mils (50 μm) or more in thickness, a 50% wastage of the coating is inevitable. Under such circumstance, the effective transfer efficiency is only 50%.

(4) **Uniformity of Applied Film Thickness.** A flat, fan-shaped spray pattern can hold film thickness variations within the 10% acceptable range, in a well-designed and engineered finish application system. The circular doughnut-shaped spray pattern is preferred in many spray applications, as it delivers film thicknesses in the range of 1 mil (Fig. 8.10).

Fig. 8.10 Correct spraying technique and spray pattern

(5) **Edge Build-Up.** In electrostatic spraying, the edges or shaped parts of the earthed workpiece tend to attract the coating spray that would ordinarily pass by the workpiece. Therefore, coating builds up on the edges, which may be considered a waste (Papadopoulos and Taghiyari 2019). Such a build-up of coating on the edges must be sanded down and touched-up manually to ensure quality finish.

(6) **Need for Manual Touch-Up.** Further, in electrostatic coating application, the electrostatic field forces may hinder the coating particles from covering the recessed parts. Therefore, it is a normal practice to overcoat, or manual touch-up such areas, which in turn lowers the transfer efficiency even further.

8.8 Optimizing Finish Application

There are several different techniques available to furniture manufacturers for the application of finishes/coatings for furniture. Each coating application technique has its advantages and disadvantages, but some system may offer specific variations that make it suitable for applications (Schaller and Rogez 2007; Yaremchuk et al. 2016). Hence, it is imperative for the furniture manufacturer to consider budgetary and logistics constraints, before choosing the most suitable coating application technique for his/her needs. Therefore, in setting up a finishing/coating section in the furniture factory, the following factors must be thoroughly considered to ensure a productive yet economical operation:

- Substitute solvent-borne coatings with low-VOC coatings.
- Achieving high transfer efficiency during spraying.
- Well-trained operators to handle the different spraying equipment effectively.
- Improve housekeeping, maintenance, and work practices in the finishing section.

- Use a heater to adjust and maintain the viscosity of the coatings used.
- Set application standards for all operations in the finishing section.

In recent years, the developments in nano-technology have impacted the global coating industry. Since last decade, research reports on the addition of nano-particles into wood coatings formulation and its effect on performance have been extensively published (Grüneberger et al. 2014; Papadopoulos and Taghiyari 2019). These developments have revolutionized the global coating industry, as coatings with specific film properties could be developed at lower cost, and higher performance levels. In fact, nano-technology also allows the formulation of low-emission coatings which are more environmentally complaint.

In the final analysis, however, the finishes/coatings application in furniture manufacturing must be carried out to ensure highest quality of furniture finish, at the lowest cost possible.

Summary

- Finishing or coating imparts the aesthetic appeal to the furniture, apart from providing it with protection against minor blemishes.
- It is a cosmetic operation, of the highest value-adding process in furniture manufacturing.
- Finishing or coating can account for almost 12% of the total furniture manufacturing cost.
- A good furniture finishing operation involves the selection of the most suitable finish material, applied through the appropriate technique in multiple steps, and allowed to cure sufficiently.

References

Bulian F, Graystone J (2009) Wood coatings—theory and practice. Elsevier B.V., Oxford, United Kingdom

Bentley J, Turner GPA (1998) Introduction to paint chemistry and principles of paint technology, 4th edn. Chapman & Hall, London, United Kingdom

Charron A (1996) Spray finishing. The Taunton Press, Newtown, Cincinnati, United States of America

Csanády E, Kovács Z, Magoss E et al (2019) Optimum design and manufacture of wood products. Springer International Publishing, Cham, Switzerland

Flexner B (2010) Understanding wood finishing: How to select and apply the right finish. Fox Chapel Publishing, Mount Joy, Pennsylvania, United States of America

Grüneberger F, Künniger T, Zimmermann T et al (2014) Nano-fibrillated cellulose in wood coatings: mechanical properties of free composite films. J Mater Sci 49:6437–6448

Hamburg HR, Morgan WM (1979) Hess's paint film defects—their causes and cure, 3rd edn. Chapman & Hall Ltd., London, United Kingdom

Hse CY, Kuo ML (1988) Influence of extractives on wood gluing and finishing—a review. Forest Prod J 38(1):52–56

Jones FN, Nichols ME, Pappas SP (2017) Organic coatings: science and technology. Wiley, Hobeken, New Jersey

Jocham C, Schmidt TW, Wuzella G et al (2011) Adhesion improvement of powder coating on medium density fibreboard (MDF) by thermal pre-treatment. J Adhes Sci Technol 25(15):1937–1946

Kúdela J, Liptáková E (2006) Adhesion of coating materials to wood. J Adhes Sci Technol 20(8):875–895

Kieblesz P, Rożanska A, Korycinski W (2017) The influence of traditional wood finishing substances on surface roughness. Annals of Warsaw University of Life Sciences—Forestry and Wood Technology No. 99. 5–14

Prieto J, Kiene J (2018) Wood coatings. Vincentz Network, Hannover, Germany

Papadopoulos AN, Taghiyari HR (2019) Innovative wood surface treatments based on nanotechnology. Coatings 9(12):866–878

Ratnasingam J, Scholz F (2006) Optimal surface roughness for high-quality finish on Rubberwood (Hevea brasiliensis). Holz Roh Werkst 64(4):343–345

Ratnasingam J, Ioras F (2013) Finishing characteristics of heat treated and compressed Rubberwood. Eur J Wood Wood Prod 71(1):135–137

Schaller C, Rogez D (2007) New approaches in wood coating stabilization. J Coat Technol Res 4(4):401–409

Slabejová S, Šmidriaková M, Fekiač J (2016) Gloss of transparent coating on beech wood surface. Acta Fac Xylologiae Zvolen 58(2):37–44

Suleman YH, Rashid SH (2007) Chemical treatment to improve wood finishing. Wood Fiber Sci 31(3):300–305

Talbert R (2008) Paint technology handbook. CRC Press, Boca Raton, Florida, United States of America

Tank GF (1991) Industrial paint finishing techniques and processes. Ellis Horwood Ltd., West Sussex, United Kingdom

Tichenor BA, Guo Z (1991) The effect of ventilation on emission rates of wood finishing materials. Environ Int 17(4):317–323

Tichy RJ (1997) Interior wood finishing—industrial use guide. Forest Product Society Publication, Madison, Wisconsin, USA

Tsotsas E, Mujumdar AS (2011) Modern drying technology—energy savings (Volume 4). John Wiley-VCH Verlag, Hamburg, Germany

Ugulina B, Hernândez RE (2018) Analysis of sanding parameters on surface properties and coating performance of red oak wood. Wood Mater Sci Eng 13(2):64–72

Yaremchuk L, Olyanyshen T, Hogaboam L (2016) Selection of coating materials for wood finishing based on a hierarchical analysis method of their technological, economic, and ecological criteria. Int J Energ Technol Policy 12(3):14–22

Chapter 9
Upholsteries For Furniture

Upholstery refers to the application of cushion, padding, and cover material to a furniture frame, to make it aesthetically more appealing and increasing comfort. It is a traditional trade and is still a labor-intensive operation in many parts of the world. Several different materials, both natural and synthetic, are being used and applied for upholstery purpose. With a growing concern for environmental and safety compliance, regulations related to its strength, durability, and fire resistance have been mandated in many markets for upholstered furniture.

9.1 Introduction

The term upholstery refers to attaching furniture frames with cushioning, padding, and textured material covers. Traditionally, upholstery is a skilled job where workers with experience and in-depth skills are hired by factories to work on leather, fabric, wooden frames, steel designs, etc. to give a perfectly fitted upholstery to furniture (Dangelico et al. 2013; James 2016).

Traditional upholstery has a long history. It is a skill that has evolved over centuries, from simply padding and covering *chairs*, seats, and *sofas*, to incorporating springs to seats and sofas, making it more comfortable. The application of lashings, stuffing of wools, hessians, scrims, bridle ties, and stuffing ties made of animal hair, grasses, and coir, sewing, top stitching, flocking, and wadding were all done by hand, which reflect the developments in the field of upholstery over the centuries (Cooke 1987).

In the present day, modern furniture upholstery is more likely to be made entirely or partially of cellular polyurethane (PU) foam. Structure, resilience (load recovery), and, most importantly, lightweight is provided by such materials. The outer decorative textile, or fabric, is then applied. Owing to the fact that these synthetic polymers will age and lose their performance, such materials must be carefully selected to fulfill its intended application (Fulton and Weston 2017). The importance of high-quality

J. Ratnasingam, *Furniture Manufacturing*, Design Science and Innovation,
https://doi.org/10.1007/978-981-16-9412-7_9

foam cannot be underplayed in upholstery. The crisp and linear design is timeless and very versatile, and with the choice of an appropriately specified foam that would maintain its shape over time, without sacrificing the comfort. Further, as upholstery is combustible, conformance to furniture fire regulations (FFR) is mandatory in almost all jurisdictions throughout the world (Krasnys et al. 2000).

9.2 Foam for Upholstered Furniture

A variety of foam ranging from expanded polystyrene (EPS), polyethylene (PE), polyurethane (PU) foam, etc. are commonly found in the furniture industry, but the PU foam has the largest market share. The density of foam reflects the mass per unit volume, and it can range from 32 kg/m^3 to 240 $^{kg}/m^3$. High-density foams are optimal for uses that get overwhelming application daily, such as lounge chair pads, bedding, or car seating. Firmness (i.e., compression), interprets the feel of foam and how it yields to weight and pressure. Its estimation is called Indentation Load Deflection (ILD), which is ascertained by mechanical performance testing (Xu et al. 2015). A *hard foam* material will require more force to reach 25 percent compression, and a gentle material will require less force. Most common foams have ILD values from 8 to 70, with some foams being as high as 120 to 150. Firmness testing is done to assist in classifying the form according to its ability to bear weight in end-use applications, i.e., whether it is for seating, sofa, or the chair back. It must be emphasized that for most furniture upholstery applications, the expected lifespan of a foam cushion is primarily dependent on the density and thickness of the foam, but it is generally about 5–7 years (Fulton and Weston 2017). If the foam is of the high resilience type, it tends to recover its shape faster and better after use.

9.3 Fabric

For a start, fabric makes an enduring impression of the furniture, so it is important that it is attractive and comfortable to the feel. When selecting upholstery fabrics for furniture, considerations must be given to how simple it is to upholster, the nature of the pile/weave, its wear resistance, and fire retardancy (Gandhi and Spivak 1994). All these components influence how the fabric will perform over time, particularly when forms and curves are thrown into the blend or wide swathes of fabric.

The Martindale method is a method to evaluate the natural abrasion or wear of the fabric. In this method, the fabric is brushed against a grating surface with an indicated constraint. The higher the wearability, the more scrap resistant the fabric is. For most furniture applications, there is a rule of 30,000 Martindale minimum (Yates 2002).

One of the foremost criteria when it comes to material determination for furniture is how much utilization it can stand up to. This can be measured in twofold rubs, which is also known as the Wyzenbeek method (Yates 2002). If the material is between 8,000 and 10,000 twofold rubs, it is planned for light utilization, and it will not support regular wear and tear. Medium utilization upholstery material is between 10,000 and 15,000 twofold rubs. Most private upholstery material falls into this category and can withstand 10,000 to 25,000 double rubs. Although the *Martindale* and *Wyzenbeek* methods are both textile abrasion tests, which test is required in the market is dependent on the standard applied (Warfield 1987).

Another important measure is the thread count in woven fabrics. Thread count refers to the number of threads woven together in a square inch (O'Neill 1999). A thread count of 150 (75 strings one way, 75 the other) produces muslin, which feels a little harsh, certainly not luxurious. Thread count is the trade 'buzz word' for promoting extravagance sheets, shirts, and other woven fabric goods. Generally, better strings can weave together easily, producing softer and better fabric but at a higher price tag. Good quality fabrics generally come in at 180, and anything above 200 is considered superior quality.

There are many types of fabrics available to upholstery application in furniture (Table 9.1), and a detailed discussion on the various types of fabrics is provided by Yates (2002).

Table 9.1 Fabrics for upholstery

Linen	Made from flax, linen is an exceptionally sturdy natural fiber. It is smooth, soft, and is frequently used in a mix with cotton for greater elasticity
Cotton	Cotton is soft and durable but is likewise prone to wrinkling and may be easily soiled. It common for slipcovers, considering that most cotton may be clean with cleaning soap and water. Cotton upholstery fabric is regularly a part of a mix, containing approximately 45% to 60% cotton
Wool	Wool is a natural fiber that comes from animal hair and is highly desirable as upholstery fabric. However, it may be a scratchy, and if not blended, may be difficult to clean and risk felting
Leather	Made from animal cover, leather is durable and easy to clean. However, there are natural, artificial, and mixed leathers in the market, which may be considered for upholstery
Polyester	Polyester is an overall performance synthetic cloth that is used in tandem with natural substances, such as cotton and wool. Polyester blends offer strength, easy cleaning, and resistance to fading, wrinkling, and abrasion
Microfiber	Microfiber is a knit blend polyester material, which is softer than suede and easier to clean. It is product of tightly woven artificial fibers that offer sturdiness and moisture resistance
Rayon	Rayon is a cellulose-based material and changed into advanced as an imitation for silk, linen, and cotton. Although it is durable, it is vulnerable to wrinkling. It is generally mixed with other materials for use in upholstery
Sunbrella	Sunbrella is an acrylic fabric that is commonly used for outdoor furniture upholstery. It is resistant to environmental factors

*Source*Yates (2002)

When joining fabric to complicated shapes and structures, distinctive sewing procedures are employed. An example is the double topstitch (Happlewhite 2012). This gives the upholstery additional quality and control (expanding the product's longevity). It moreover complements the compact shape of the couch or sofa. Nevertheless, it must be emphasized that the fabric makes a noteworthy distinction to the visual personality of the chair, whereas the inside components may not be obvious, they make a huge distinction to the consolation and ergonomics of the piece.

9.4 Leather

Leather is an important upholstery material for furniture, especially for high-end furniture (Ciupan et al. 2018). It could be a solid and adaptable fabric created by *tanning* animals *rawhide* and *skins*. The most common crude fabric is cattle skin. It can be delivered at fabricating scales ranging from artisan to cutting-edge mechanical scale. Leather is utilized to form an assortment of articles, from footwear, vehicle seats, clothing, bags, book ties, adornments, to furniture. It is created in a wide assortment of sorts and styles and brightened by wide ranging techniques (Hole and Whittaker 1971). The leather fabricating technique is separated into three principal sub-processes: preliminary stages, tanning, and crusting. Another sub-process known as wrapping up can be included into the leather handle arrangement, but not all leathers get finished in the market (Wood 2009).

Leather is a valuable material for upholstery for furniture, and has the following properties:

1. High Tensile Strength

   The tensile strength of leather is the malleable strain of leather until it tears. The malleable quality is exceptionally distinctive within the longitudinal and transverse directions of the leather skin. The average tensile strength of leather is between 8 and 25 N/mm^2, while typically furniture requires at slightest 200 N per 5 cm (Law and Gentles 2020).
2. Resistance to Tear

   Good leather is steady and safe to tearing, while suede, nubuck, or amazingly delicate lambskin will not have the same stability.
3. Flex Resistance

   The term flex resistance refers to the ability to resist various flexing cycles without harm or disintegration. Most commonly, leather items and components are harmed by repeated flexing cycles within the shape of surface cracks.
4. Resistance to Puncture

   Puncture resistance signifies the relative capacity of leather to repress the interruption of a foreign object. Cut safe leather is usually secured from punctures, cuts, tearing, and abrasions.

5. Heat Insulation

   Leather contains a great deal of air, which is a destitute conductor of heat. It could be great heat obstruction and gives fabulous heat insulation.
6. Permeability to Water Vapor

   Leather will hold a high amount of water vapor. This property empowers leather to retain sweat, which is afterward dissipated, providing comfort (Namicev and Tasevska 2019).
7. Thermostatic Properties

   Leather is warm in winter and cool in summer and is additionally safe to heat and fire.
8. Moldability

   Leather can be molded and will hold its new shape.
9. Resistance to Wet and Dry Abrasion

   Leather can resist wear in both damp and dry situations.
10. Resistance to Biodegradation

    Leather is resistant to attack by biodegrading organisms, such as fungi and mold.
11. Resistance to Chemicals

    Leathers are generally tanned and dressed to resist chemicals.

In later times, however, *faux leather*, also known as *synthetic leather*, is a petroleum-based alternative leather which is making inroads into the upholstered furniture market. In many instances, it has comparable and alluring properties of natural leather. Like genuine leather, faux leather is delicate to the touch, and it is water resistant. In this manner, this material is highly resistant to stains, and it is simple to clean. Although, synthetic leather is less durable than genuine leather, its resistance to wear and tear apart from its lower cost makes it a preferred alternative material for upholstered furniture (Fulton and Weston 2017).

Against this setting, upholstery furniture producers must have a great understanding of the materials accessible within the market, to stay competitive, while ensuring fulfillment to their clients through quality upholstered furniture. Research by Zi and Bullard (2008, 2009) has emphasized that upholstery furniture manufacturing around the world is reeling from competitive pressures from the lower cost manufacturing nations. Upholstered furniture manufacturing remains labor intensive, and thus, given the extent of value addition within the product range and market, manufacturing such furniture competitively will continue to be a challenge (Ratnasingam 2016). From another perspective, innovation and value addition within the upholstered furniture industry must be improved, to pave the way for an economically rewarding industry (Rapôso et al. 2021). *Greater application of information and computing technology (ICT) and automation should be encouraged within the industry to boost productivity* (Staneva et al. 2018). In this manner, by expanding the stylish offer and appeal of upholstered furniture, its manufacturing will emerge competitive and financially rewarding.

Summary

- Upholstery is a skilled and value-adding process to furniture.
- The selection of the suitable foam and fabric is important to ensure the aesthetic appeal of furniture.
- Upholstery must also comply with the furniture fire regulations.

References

Ciupan E, Ciupan Câmpean EM, Stelea L et al (2018) Opportunities of sustainable development of the industry of upholstered furniture in Romania a case study. Sustainability 10(9):3356

Cooke Jr, ES (1987) Upholstery in America and Europe from the Seventeenth Century to World War I. Barra Foundation, Wayne, Pennsylvania, United States of America

Dangelico RM, Pontrandolfo P, Pujari D (2013) Developing sustainable new products in the textile and upholstered furniture industries: role of external integrative capabilities. J Prod Innovat Manag 30(4):642–658

Fulton N, Weston S (2017) The Upholsterer's handbook. Octopus Publishing Ltd., London, United Kingdom

Gandhi S, Spivak SM (1994) A survey of upholstered furniture fabrics and implications for furniture flammability. J Fire Sci 12(3):284–312

Happlewhite G (2012) The cabinet maker and upholsterer's guide, 3rd edn. Dover Publication, New York, United States of America

Hole LG, Whittaker RE (1971) Structure and properties of natural and artificial leathers. J Mater Sci 6:1–15

James D (2016) Upholstery: a complete course. Guild of Master Craftsman Publication, East Sussex, England

Krasnys J, Parker W, Babrauskas V (2000) Fire behaviour of upholstered furniture and mattresses. Noyes Publications, Park Ridge, New Jersey, USA

Law A, Gentles P (2020) A beginner's guide to upholstery. CICO Books, Holborn, London, United Kingdom

Namicev P, Tasevska V (2019) Specifications of eco-materials and their influence in the design of modern furniture. J Process Manag New Technol 7(1):12–20

O'Neill H (1999) Upholstery—a manual of techniques. Crowood Press Ltd., Wiltshire, United Kingdom

Ratnasingam J (2016) Upholstered furniture manufacturing—a guide for entrepreneurs. Tech. Note. No. 11, IFRG Publication, Singapore

Rapôso A, César SF, Kiperstok A (2012) Cleaner production and life cycle design of upholstered furniture. Int J Environ Sustain Dev 11(3):217–237

Staneva N, Genchev Y, Hristodorova D (2018) Approach to designing an upholstered furniture frame by the finite element method. Acta Facultatis Xylologiae Zvolen 60(2):61–69

Warfield CL (1987) Upholstered furniture—results of a consumer wear study. Text Res J 57(4):192–199

Wood L (2009) Upholstered furniture in the lady lever art gallery. Yale University Press, New Haven, Connecticut, USA

Xu W, Wu Z, Zhang J (2015) Compressive creep and recovery behaviours of seat cushions in upholstered furniture. Wood Fibre Sci 47(4):431–444

Yates M (2002) Fabrics—a guide for interior designers and architects. W.W. Norton & Company, New York, USA

Zi W, Bullard SH (2008) Firm size and competitive advantage in the U.S. upholstered, wood household furniture industry. Forest Prod J 58(1/2):91–96

Zi W, Bullard SH (2009) Competitive strategy and business performance in the U.S. upholstered, wood household furniture industry. Forest Prod J 59(9):15–19

Chapter 10
Furniture Packaging

Packaging is the science, art, and innovation of securing a product for handling, delivery, and utilization. Apart from that, packaging is also important to entice potential customers to purchase the product. In the case of furniture, packaging is mostly accomplished with the use of corrugated carton box as the exterior packaging, while the interior packaging uses materials such as bubble wrap, polyethylene sheet and foam, and polystyrene. Due to the weight of furniture, packaging of furniture must be carried out with utmost care, to ensure no damages occur during handling and transportation. The packages are usually subjected to drop tests to assess its quality, suitability, and sufficiency of protection.

10.1 Introduction

Packaging is the science, art, and innovation of encasing or securing items for conveyance, handling, and utilization (Anon 1995). It may be a facilitated framework of planning merchandise for transport, warehousing, coordination, dealing, and end use (Fig. 10.1). In fact, packaging also has a strong effect on alluring clients to purchase the item (Raheem et al. 2014). It has been detailed by Ratnasingam (2019) that packaging can contribute up to 8% of the total materials cost of the furniture, and its financial orientation cannot be underplayed.

Generally, packaging serves several purposes, which incorporate, physical assurance (i.e., against harm), obstruction assurance (i.e., against the climate components), control or agglomeration (i.e., gathering little objects together), data transmission (i.e., how to utilize), showing (i.e., alluring customers), security (i.e., decreasing damage), comfort (i.e., ease of conveyance), and brand situating (i.e., publicizing and picture building). For an extensive discussion of packaging, the readers are referred to Emblem (2012), Twede et al. (2015), and Szaky (2019).

In the case of furniture, packaging plays an imperative part of ensuring the furniture is protected from harm and damage during transit. Being a merchandise and

J. Ratnasingam, *Furniture Manufacturing*, Design Science and Innovation,
https://doi.org/10.1007/978-981-16-9412-7_10

Fig. 10.1 Furniture packaging

fashion, the furniture must reach the customer's site in flawless condition (Hughes 1996).

A well-packaged furniture must be able to preserve its imperative characteristics, during transportation and storage of the item. The packaging must be clearly recognizable, and provide sufficient assembly and utilization instruction, apart from highlighting any potential risk notices, through pictures and illustrations for easy understanding (Csanády et al. 2019). In this manner, furniture packaging must be sturdy to maintain the quality of the item, as well as meeting the customer's inclinations.

Furniture packaging is also an important critical publicizing medium to attract potential customers. In this context, with the advancements in packaging design, numerous packaging points are taken into consideration, such as color, package design, printed information, and the innovative packaging methods to generate the proper enthusiastic discernment among the shoppers (Klevås 2005).

10.2 Packaging Materials

Several different materials are used for packaging, ranging from folded sheets and cartons, cellophane film, polyethylene (PE) foam, polystyrene (*Styrofoam*), Bubble Wrap, cotton, plastic sheet, paper, paper adhesive tapes, and polypropylene (PP) strapping (Soroka 2009). A few of these packaging materials are pictured in Fig. 10.2.

Corrugated boards and cartons are the most common outside packaging material for furniture. It is made from one or more sheets of curved paper known as 'fluting' secured by an adhesive to two or more liners. It may have single, twofold, or triple

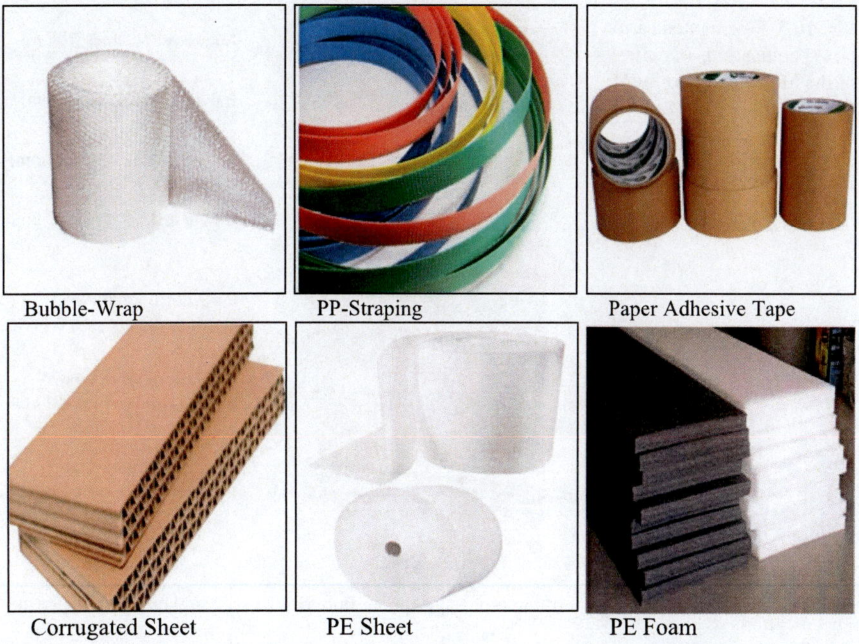

Fig. 10.2 Common furniture packaging materials (Courtesy of the Malaysian Furniture Council)

divider and is made from a distinctive type of paper, known as liners (Twede et al. 2015). The crude material used in making the corrugated boards may be wood fiber, bagasse, rice husk, bamboo, etc. The liners are ordinarily made from recycled fiber sources of both short and long fibers (Fig. 10.3). Corrugated cartons come in different sizes and thickness, to provide various degrees of protection.

The type and number of flutings impacts the degree of protection offered by the corrugated carton box (Table 10.1). Separated from rendering a steady pad for furniture, which minimizes the risk of damage, due to movement, packaging also keeps the furniture secure for long-distance transportation. Packaging also protects the furniture from the effect of humidity (Csanády et al. 2019).

The design (single face, single wall, double wall, triple wall), fluting type, bursting strength, edge crush strength, flat crush strength, basis weights of components (grams per square meter), surface treatments, and coatings can all be used to identify corrugated carton boxes. The corrugated medium, with the different fluting type using different adhesives, and linerboards can all be manipulated to create a corrugated carton box with specific properties that can be used in a variety of applications. (Kirwan 2005). Corrugated board with double and triple walls is also available for increased stacking strength and cut resistance.

The grade of the corrugated carton box is defined by the quality and strength of the lining papers used, which is defined by weight per square meter (GSM) as well as the content mix of new and recycled fiber used in the manufacture of the liner papers.

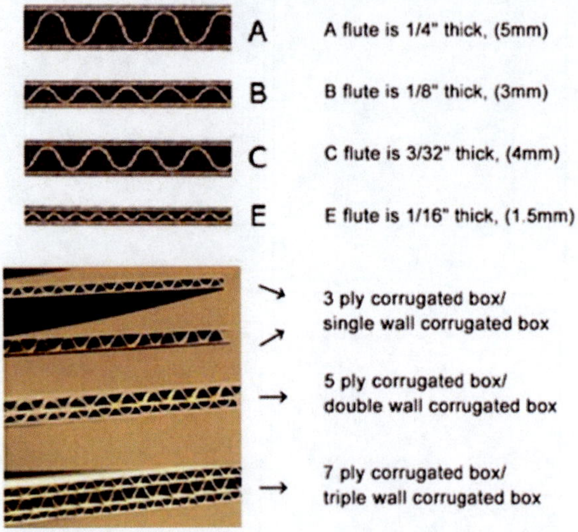

Fig. 10.3 Corrugated carton box specifications (Courtesy of the Malaysian Furniture Council)

Table 10.1 Fluting specifications used in corrugated carton

Flute type	Average flute height (mm)	Number of flutes per m
N	0.45	558 ± 15
G	0.60	587 ± 15
F	0.75	422 ± 15
E	1.15	297 ± 15
D	1.70	208 ± 15
B	2.45	155 ± 10
C	3.60	130 ± 10
A	4.65	110 ± 10
K	6.60	82 ± 7

Generally, kraft paper, ‘K’, is of better quality compared to the test paper, ‘T’. As an example, the grade 125 K/T B would mean the weight of the paper used is 125 g per square meter, while the external paper being Kraft and the internal being test quality with ‘B’ fluting in between. Further, by varying the parameters, it is possible to manufacture carton boxes with specific properties. For instance, the B or E flutes, with 125 K/T (which reflects the grammage of the kraft and test liner used), will carry a product weighing 4–6 kg. On the other hand, double-wall BC or EB flutes, with 300 K/T, will support a product weighing between 40 and 45 kg (Kirwan 2005). The weight of the paper used in corrugated cartons is expressed in grammage and is detailed in g/m^2. Among the most common grammage found in the market are 125 GSM, 150 GSM, 185, 200 GSM, and 300 GSM. The ideal humidity for corrugated carton is between 6.5 and 9.5%. In case the humidity is below the optimum point, it will cause the corrugated board to break, and if it is above that limit, it will reduce the board's compression strength (Twede et al. 2015).

The edge crushing strength (kN/m) reflects the force per unit width, which in turn predicts the compression strength of the box. The burst strength (N/mm^2) is the pressure required to rupture a corrugated sheet. The box compression strength (N) is a direct measurement of the performance of the corrugated box. The flat crush strength (N/mm^2) indicates the rigidity of the flutes (Kirwan 2005). Bending resistance reflects the force necessary to bend a rectangular paperboard through a 15° angle. Tear resistance is the force required to tear a carton board sheet along an existing initial tear, or incision. Being anisotropic, the properties of corrugated boards are usually directional in nature. Hence, many of the strength properties and surface characteristics are impacted by the orientation of the flutes and the machine direction during manufacture.

Once the specification of the corrugated box is completed, a box certificate is printed on an exterior bottom surface, with a few details about the box's strength (Twede 1992). Among the details included are bursting strength or edge crush strength, size, and gross weight limit (Wever 2011). By and large, corrugated cartons with the K-type fluting with a double-wall specification is commonly used in the furniture industry (Zhang 2011).

There are numerous designs of corrugated cartons, and each serves a specific product and customer inclination. It must be recognized that corrugated carton design and the packaging industry is a multi-billion dollars global industry (Hanlon and Kelsey 1998). The most common packaging designs used for furniture are as shown in Fig. 10.4.

On the other hand, the most common internal packaging materials found in the market are polyethylene-foam cushioning, Bubble Wraps, plastic, and paper sheets, which are less stiff, with lower density, strength, and hardness (Hughes 1996). This will guarantee that the furniture has low risk of any harm or damage from the internal packaging. Although plastic sheets are less impacted by environment factors, particularly temperature and humidity changes, it should be dry before being utilized as internal packaging materials to avoid any micro-organism (i.e., mold) growth. All things considered, the application of plastic sheets is less preferred due to its negative effect on the environment (Meherishi et al. 2019).

Once the items are packaged, they are stacked on a skid or pallet, before it is shipped to the client, or until an order is received at the warehouse. It must be recognized that furniture packaging has somewhat not been a focus of packaging research previously (Nethercote 1974), but with growing worldwide concern for sustainability, furniture packaging is attracting increasing research interests (Twede et al. 2000).

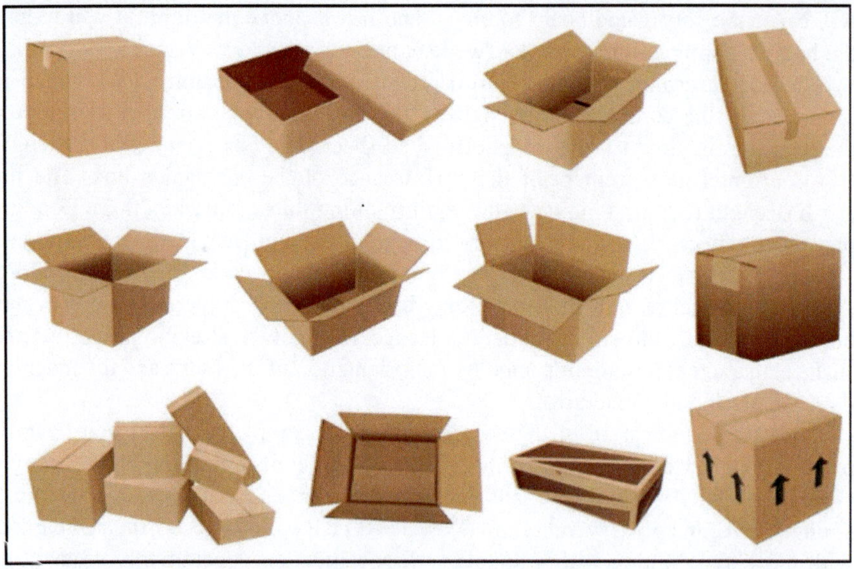

Fig. 10.4 Packaging design for furniture (Courtesy of the Malaysian Furniture Council)

10.3 Packaging Quality

In furniture packaging, there is no one-size-fits-all solution which may be regarded as the standardized method. This is because every piece of furniture has diverse shapes and measurements, and thus requires diverse packaging (Csanády et al. 2019). In fact, there is no such thing as 'too much packaging' in any circumstances, but it must be financially practical for the furniture producer. Therefore, furniture packaging is done to the best to avoid any customer disappointment. In practice, the quality of furniture packaging is ascertained through the 'drop test'. This test follows the International Safe Transit Association (ISTA) conventions (Anon 2018), which stipulates distinctive drop points and the sequence of the drop test (Fig. 10.5). In many instances, furniture buyers would specify the number of drop tests to be conducted as stipulated by the ISTA.

Furniture packages comes in various sizes and weights, and so the ISTA conventions stipulate the heights of the drop test depending on the weight of the package (Tables 10.2 and 10.3).

Once the packaged furniture passes the drop test, the packaging instruction is followed closely for all such items manufactured in the factory (Connolly et al. 2003). If the package did not pass the drop test (i.e., furniture is harmed), the packaging is strengthened further, and the drop test is conducted again until it passes. In most furniture factories, the drop test is conducted when the furniture buyer or their representative is present at the factory premise (Anon 2020).

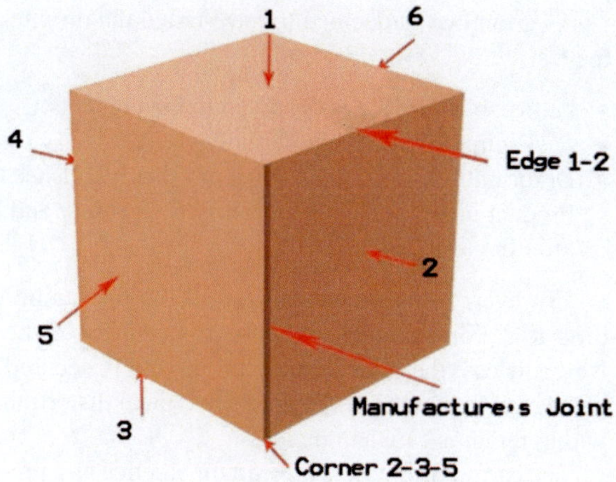

Fig. 10.5 Drop test directions

Table 10.2 Drop test protocols

Sequence	Orientation	Specific face, edge, or corner
1	Corner	Most fragile face-3 corner, if not known, text 2–3–5
2	Edge	Shortest edge radiating from the corner tested
3	Edge	Next longest edge radiating from the corner tested
4	Edge	Longest edge radiating from the corner tested
5	Face	One of the smallest faces
6	Face	Opposite small face
7	Face	One of the medium faces
8	Face	Opposite medium face
9	Face	One of the largest faces
10	Face	Opposite large face

Source ISTA

Table 10.3 Drop test heights

Package weight (kg)	Drop height (cm)
1–9	76
10–18	61
19–27	46
28–45	30
46–68	20

Source ISTA

A carton box is deemed to have failed the drop test if the following occurs after the test:

- Noticeable damage is observed to the contents.
- Packaging is damaged internally.
- Deformation, scratches, dents, or other blemishes to the items inside.
- Product usefulness is compromised, or safety and health risks posed by the item after the test (Anon 2020).

However, some minor damage to the packaging cannot be avoided during the drop test. For example, the corners or edges of the carton might indent slightly. Nevertheless, the item should still be totally secured within the package. A carton passes the test, when there are only typical distortions to the carton, while the item within remains safe and intact.

The carton drop test is a common practice as a pre-shipment assessment (Hughes 1996; Anon 2020). Some furniture manufacturers would conduct drop tests prior to the final inspection to evaluate the packaging quality before the arrival of the furniture buyer or the representative. Further, records of the results from each drop test should be kept for future improvement.

It must be recognized that packaging quality control is simply about delivering a product in impeccable condition to the customer, and not delivering a damaged, non-acceptable product. In this context, the carton drop test is just one of the numerous on-site packaging checks that is performed to assess packaging quality (Emblem 2012). Checks for inferior packaging and compliance with packaging standards provides the necessary assurances that the packaging meets the requirements.

10.4 Packaging Machines

The common packaging machines found in the furniture industry are shown in Fig. 10.6. The criteria taken into consideration when choosing the appropriate packaging machine depend on its required specialized capability, manpower needs, productivity, user-friendly, adaptability, and, of course, the investment cost (Csanády et al. 2019).

Despite the accessibility to these packaging machines, in most furniture factories in the developing world, the packaging operation is often a semi-automatic operation, with a sizeable number of workers (Wang et al. 2021). On a global scale, the packaging industry is embracing sustainability and moving towards becoming a green industry (Coelho et al. 2020; Wandosell et al. 2021). In line with the push for the circular economy, the packaging industry is expanding the 3R concept, i.e., *reuse, recycle,* and *reduce*, to the next level of becoming sustainable (Szaky 2019). Nevertheless, it must be recognized that change in the furniture industry takes time to root, but with growing external forces, the industry is poised to meet its global obligations. In any case, packaging remains an operation that furniture manufacturers cannot bear to downplay, as it serves as the first point of contact with the potential customer.

Fig. 10.6 Packaging machines (Courtesy of the Malaysian Furniture Council)

Summary

- Packaging not only provides protection to the furniture but it also serves many other secondary purposes.
- Corrugated carton box is the most common external packaging material used in furniture packaging.
- Internal packaging materials used include bubble wrap, PE foam, paper, and plastic sheets.
- The package is then sealed using paper adhesives and PP strappings.
- The quality of the packages is ascertained through the ISTA sanctioned drop test.

References

Anon (1995) Manual on the packaging of furniture. International Trade Centre/UNCTAD/GATT Publication, Geneva, Switzerland

Anon (2018) Packaging testing & guidelines. International Safe Transit Association (ISTA), East Lansing, Michigan, USA

Anon (2020). Furniture packaging guidelines. Crate & Barrel Company, Northbrook, Illinois, USA

Coelho PM, Corona B, Klooster RT et al (2020) Sustainability of reusable packaging—current situation and trends. Resour Conserv Recy: X 6: 100037

Connolly SM, Marcondes JA, Weigel TG et al (2003) A comparison of the performance tests used for furniture packaging. J Test Eval 31(3):253–260

Csanády E, Kovács Z, Magoss E et al (2019) Optimum design and manufacture of wood products. Springer International Publication, Cham, Switzerland

Emblem A (2012) Packaging technology—fundamentals, materials and processes. Woodhead Publishing, Cambridge, UK

Hanlon JF, Kelsey RJ (1998) Handbook of package engineering, 3rd edn. CRC Press Boca Raton, Florida, USA

Hughes P (1996) Exporting furniture: getting the packaging right. Int Trade Forum 1:1–8

Kirwan MJ (2005) Paper and paperboard packaging technology. Blackwell Publishing Ltd., Oxford, UK

Klevås J (2005) Organization of packaging resources at a product-developing company. Int J Phys Distr Log 35(2):116–131

Meherishi L, Narayana SA, Ranjani KS (2019) Sustainable packaging for supply-chain management in the circular economy—a review. J Clean Prod 237: 117582

Nethercote CH (1974) A review of current problems and research in furniture packaging. Eastern Forest Products Laboratory, Quebec, Canada

Raheem AR, Vishu P, Ahmed AM (2014) Impact of product packaging on consumer's buying behaviour. Eur J Sci Res 122(2):125–134

Ratnasingam J (2019) Furniture packaging—a review. Tech. Note, no 4. IFRG Publication, Singapore

Soroka W (2009) Fundamentals of packaging technology, 4th edn. Institute of Packaging Professionals, Napperville, Illinois, USA

Szaky T (2019) The future of packaging—from linear to circular. Berrett-Koehler Publishing, Oakland, California, USA

Twede D (1992) The process of logistical packaging innovation. J Bus Logist 13(1):69–94

Twede D, Clarke RH Tait JA (2000) Packaging postponement: a global packaging strategy. Packag Technol Sci 13(3): 105–115

Twede D, Selke SEM, Kamdem D-P et al (2015) Cartons, crates and corrugated board. Handbook of paper and wood packaging technology. DEStech Publication Inc., Lancaster, Pennsylvania, USA

Wandosell G, Parra-Meroño MC, Alcayde A et al (2021) Green packaging from consumer and business perspectives. Sustainability 13: 1356

Wang G, Zhu J, Cai W et al (2021) Research on packaging optimization in customized panel furniture enterprises. BioResources 16(1):1186–1206

Wever R (2011) Design for volume optimization of packaging for durable goods. Packag Technol Sci 24(4):211–222

Zhang HY (2011) Rational consideration on package design of wooden furniture. Adv Mater Res 211(212):250–253

Chapter 11
Standardization and Environmental Compliance

Furniture as a utility product which must comply with market requirements. In this context, furniture manufacturing facilities are usually compliant to several standards to give credence to their compliance to market requirements. Likewise, the health and safety aspects of the work environment in the furniture industry is subjected to scrutiny to ensure that the workers' well-being is considered always. As the global market grows more in environmental compliance, in line with the Sustainable Development Goals, green manufacturing practices, which include product design, materials, and technologies, are increasingly becoming environment friendly, as the market demand for green furniture increases.

11.1 Introduction

Furniture manufacturing with its wide network of supply chain is a global industry. Perceived as a fashion and merchandize, furniture is often closely examined by potential customers before it is purchased to meet their functional, aesthetical, socioeconomic status, feel-good factor, and, of course, the ensuing pride of ownership (Saunders 2014; Xu et al. 2020). As countries around the world work towards complying with the Sustainable Development Goals (SDGs), a concept developed by the United Nations, the furniture manufacturing industry is expected to follow suit, embracing the concepts of green, cleaner industry embedded within the circular economy (Xiong et al. 2020).

Being a traditional industry, furniture manufacturing is labor intensive in nature, but it is also becoming a high energy consuming industry. Energy consumption, measured in energy consumed per unit finished product, in the furniture manufacturing industry is dependent on several factors. These include the manufacturing processes employed, state of technology, materials used, production system applied, and the type of furniture produced. A report by the International Energy Agency

J. Ratnasingam, *Furniture Manufacturing*, Design Science and Innovation,
https://doi.org/10.1007/978-981-16-9412-7_11

(IEA) in 2018 stated that the average energy consumption by the furniture manufacturing industry in the European Union was 0.0332 toe/t of finished product, as opposed to the average 0.0391toe/t of finished product in Asia (Xu et al. 2020). On this account, it is apparent that energy consumption within the furniture manufacturing industry in Asia is markedly higher, suggesting a need for greater efforts to reduce energy consumption in this part of the world (Daian and Ozarska 2009).

Although energy in the furniture industry is regarded as fixed overhead costs, it is a cost element that can be properly managed. Since energy prices worldwide is steadily increasing, furniture manufacturers throughout the world recognize that it is a valuable commodity, and must be used efficiently to ensure a positive impact on the company's profits. Consequently, many manufacturers are adopting energy-saving measures in many energy-demanding operations in the factory, such as kiln drying, heating, exhaust system, air compression, as well as the boiler operations (Cinar and Erdogdu 2018). It has been shown that energy costs can account up to 17% of the total manufacturing cost of high-end furniture, especially those using specialized coating materials (Ratnasingam et al. 2013). Against this backdrop, energy management is becoming an increasingly important function in the furniture manufacturing industry.

11.2 Standards in the Furniture Industry

To assure customers that the furniture manufactured meets their expectations, manufacturers adopt a few ISO certification standards, apart from the good manufacturing practices (GMP) scheme. Among the three most common ISO certification schemes include: (1) ISO 9001, (2) ISO 14001, and (3) OSHAS 18001.

Quality Assurance: ISO 9001

Quality is a priority when buying furniture. The **ISO 9001** is an international standard which characterizes quality management standards in the furniture production outfit, and the ISO 9001 certification implies that the furniture producers have met quality guidelines so that customers know that they are buying good product (Ruddell and Stevens 1998; Ratnasingam et al. 2013).

Focus on the Environment: ISO 14001

Furniture manufacturers that are **ISO 14001** certified must demonstrate that they meet the desired environmental standards and are making dynamic enhancement improvements to their environmental commitments. On the other hand, buyers are ensured that they are buying products that have been delivered by a company that pays attention to its environment and society obligations (Ruddell and Stevens 1998).

Safety First: OHSAS 18001

OHSAS 18001 covers all issues of occupational health and safety at work. This universal standard requires furniture companies to substantiate the claim of their commitment to ensuring the well-being, health, and safety of their workers, by embedding this requirement in their management systems (Ratnasingam et al. 2012). On the other hand, for consumers, this safety is an indication of the company's sense of responsibility towards its employees.

However, not all furniture manufacturers are ISO certified. Under such circumstances, their products may still be marketed in the European market, provided it carries a *CE* mark. The *CE* marking is a conformity marking that the products comply with health, safety, and environmental protection standards of the European Union (Fet and Skaar 2006; Holopainen et al. 2015).

Another certification scheme is the *EU Ecolabel* which is another international ecolabel scheme adopted since 1992 by the *European Commission*. The EU Ecolabel is an element within the larger EU Action Plan on Sustainable Consumption and Production, and it complies with *ISO 14020 Type I* ecolabel requirements (Fet and Skaar 2006). Specialists, business, consumer groups, and environmental NGOs collaborated to develop and review the EU Ecolabel standards. The national competent bodies are responsible for managing the applications and licenses. Since the last decade, however, the European Commission developed new EU Ecolabel standards for computers, furniture, and footwear based on a set of ecological criteria, aimed at fostering Europe's transition to a circular economy, while encouraging sustainable production and consumption in the region.

Since quality conformance is a market-driven standard, other countries have also developed their own certification schemes, such as the Nordic Swan Ecolabel and the Indian Eco-mark. The *Nordic Swan Ecolabel* works to decrease the environmental impact due to products manufacturing and use. It aims to educate consumers to select products and services that are environmental friendly. On the other hand, the *Eco-mark* certification by the Indian Bureau of Standards since 1991 also aspires to make the society at large more environmental friendly. The Eco-mark is issued for products that meet a set of standards for environmental compliance (Saunders 2014).

It must however be emphasized that all the above certification schemes are not a reflection of product quality nor a certification label, but it is the manufacturer's statement that the products meet standards for health, safety, and environmental protection in a specific marketplace (Nizialek et al. 2018). Nevertheless, in the globalized world, any certification, including the ISO certification, has been found to be a compelling marketing tool for furniture manufacturers to penetrate the different marketplace. The furniture manufacturer's notoriety is within the worldwide stage.

11.3 Safety and Health

Generally, woodworking operations differ from metal-working operations in four major aspects: (1) wood is a natural material with variable quality compared to the engineered metals; (2) wood is an insulator of heat, and therefore during cutting it has difficulty dissipating the heat; (3) the higher cutting speeds encountered in woodworking operations pose an inherent danger; and (4) wood cutting produces a higher amount of dust particles compared to metal cutting (Ratnasingam et al. 2016).

As a result, almost all woodworking industry, including furniture manufacturing, is typically stigmatized to be the 3D industry, i.e., dangerous, dirty, and difficult. To shed this negative perception of the overall woodworking industry, it is important that wood products manufacturers comply to *Occupational Safety and Health* (OSH) regulations. OSH is a multidisciplinary topic focused on the safety, health, and welfare of the workers at the workplace. A previous study by Ratnasingam et al. (2012) has reported that the prevailing safety climate in the furniture factory is determined by many factors, which include the nature and type of work, ergonomics, machine factor, workers' training, maintenance system in-place, factory layout, and workforce characteristics. Further, it must be emphasized that the effect of each of these factors on the prevailing safety climate is different for different factories.

One topic that is gaining importance in the furniture industry is ergonomics (Mirka 2005). **Ergonomic** and **human factor** is simply the science of 'fitting the worker to the job', i.e., or finding a proper match between worker-machine-environment. It aims to ensure the worker's capabilities is aligned with the tasks, information, and environment he/she is working in, to reduce human errors, increase productivity, and create a conducive work environment. Since furniture manufacturing involves many repetitive operations, the workers may suffer from repetitive strain injuries and other musculoskeletal problems over time.

The machines in furniture manufacturing pose safety risks to employees, especially when used improperly or without proper safeguards. Workers may be subjected to various degrees of accidents, ranging from minor lacerations to amputations and blindness. Poorly maintained machines can also cause accidents, and therefore it is important to keep a record of operating risks posed by every machine in the factory.

Further the use of *Personal Protective Equipment* (PPE) by the employees must be made mandatory to protect workers from mishaps and accidents. Employers should provide the necessary training to their workers to ensure safe operation of the machines.

The common safety risks encountered with woodworking machines include the point of operation (i.e., rotating or reciprocating movements), kickbacks, flying chips, knife/tool projection, and fire and electrical hazards (Ratnasingam et al. 2019).

Woodworking machines also pose health risks such as exposure to dust, noise, and vibrations (Ratnasingam and Scholz 2015). Some wood dust can trigger serious allergic reactions, thus impacting the health of the workers. In this context, the health risks posed by woodworking machines, which are both acute and chronic, must be clearly understood and appropriate protective measures must be taken. The primary health hazards from woodworking include: (1) dust particles, (2) noise, (3) vibration, and (4) chemical hazards (which includes coatings, adhesives, and solvent vapors) Table 11.1 presents the minimum exposure limits for the common OSH factors typically encountered in the furniture industry.

With growing call for the adoption of the Sustainable Development Goals (SDGs) (Fig. 11.1), customers worldwide are requiring furniture manufacturers to provide workers with the minimum acceptable living conditions, due to increasing plight of foreign or migrant contract workers, which has been a growing concern worldwide. For instance, the *Clean Air Regulation* and the *Workers' Minimum Standards* of

Table 11.1 Minimum requirements for OSH in the furniture industry

Permissible exposure limit (PEL) for wood dust over 8 hours on a time-weighted average (TWA) basis	5mg/m^3 of air
Permissible Exposure Limit (PEL) for noise to reach the 100% noise dose over 8 hr	85 dBA
Minimum lighting on factory floor	400 Lux or Foot-Candles
Total Volatile Organic Compound (VOCs) exposure limit over 8 hours	400 μg/m^3
Formaldehyde exposure limit over 8 hours TWA	0.75 ppm
Toluene exposure limit over 8 hours TWA	100 ppm
Xylene exposure limit over 8 hours TWA	80 ppm

Source Ratnasingam et al. (2012)

Fig. 11.1 Sustainable Development Goals (SDGs)

Housing and Amenities Act are the recent regulations requiring manufacturers to provide acceptable working and living conditions for workers (Ratnasingam et al. 2019).

Finishing or coating materials as well as the adhesives/glues are the common emitters of *volatile organic compounds* (VOCs), which upon exposure may pose serious health risks to workers in the industry (Fig. 11.2).

The VOC that is gaining international attention is formaldehyde and its emission comes from wood-based panels, coatings, or finishes and adhesives (Böhm et al. 2012; Bulian and Fragassa 2016; Tong et al. 2019). To minimize exposure, markets around the world have strict formaldehyde emission standards. Examples include the E0, E1, and E2 standards in Europe, the CARP-P1, and CARB-P2 requirements in the United States of America, and the F**, F***, and F**** standards in Japan. These standards have been formulated to ensure exposure to formaldehyde, is minimized in humans due to its proven health hazard (Table 11.2).

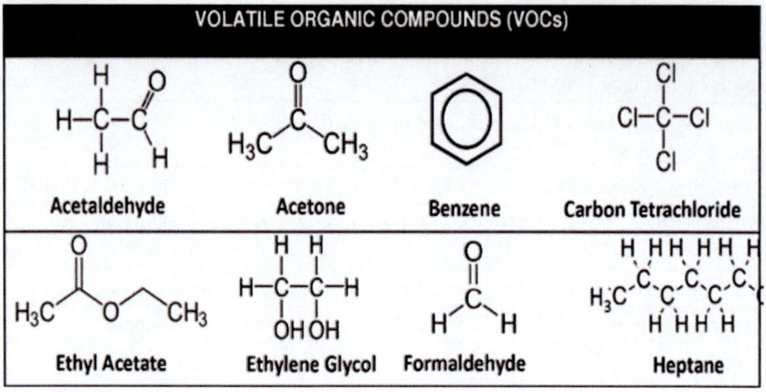

Fig. 11.2 Common volatile organic compounds in the furniture industry

Table 11.2 Common formaldehyde standards

CARB-P1: California Air Resources Board Phase 1 Emission Standard. *CARB-P2: California Air Resources Board Phase 2 Emission Standard (Enforcement date for PB & MDF was Jan 01, 2012)* *ANSI A208.1 & 2: North America voluntary standards ANSI A208.1-2009 Particleboard & ANSI A208.2-2009 MDF*	North America/United States of America
E2: European E2 Emission Standard *E1: European E1 Emission Standard* *E0: European E0 Emission Standard* *SE0: European Super E0 Emission Standard*	European Union
*F**: Japanese F-Star2 Emission Standard* *F***: Japanese F-Star3 Emission Standard* *F****: Japanese F-Star4 Emission Standard* *JIS: Japanese Industrial Standard*	Japan

Source: Tong et al. 2019

11.4 Timber Certification

As consumers around the world become increasingly environmental conscious, there is a growing market requirement for furniture manufacturers that the wood materials they use are sustainable, legal, and certified (Stevens et al. 1998; Chen et al. 2011; Espinoza et al. 2012). **Sustainable** wood originates from sustainably managed forests, and it's renewable since forest stewards manage the resources to prevent damage to eco-systems and other aspects, by taking a long perspective rather than a short one. **Legal wood** is the harvest, transportation, purchase, or sale of timber in according to the legal guidelines and rights. **Certification** is consequently meant as a 'seal of approval' to inform customers that the timber products have been produced from sustainably managed forests and its movement throughout the supply chain is documented, independently monitored by credible auditors. Certification offers an effective way to encourage corporations to enhance forest management practices, harnessing the power of markets and consumers to assist environmental and social standards. For wooden products, the **Chain of Custody** (*CoC*) certification tracks the certified material through the manufacturing, from the forests to the buyers, including all successive levels of processing, transformation, production, and distribution (Ratnasingam et al. 2008a, b). In essence, the CoC presents proof that certified wood products had used certified raw material that originated from certified forests.

The *Forest Stewardship Council* (FSC), the *Program for the Endorsement of Forest Certification* (PEFC), the *Lacey Act* (USA), the *Australian Illegal Logging Prohibition Act* (Australia), *Goho-Wood* (Japan), the *Sustainable Forest Initiative* (Canada), and the European Union Timber Regulation (European Union) are some of the most popular certification schemes and related certification legislations in the world (Fig. 11.3) (Saunders 2014).

To assist tropical countries around the world to adopt sustainable forestry practices, several global initiatives have been introduced. Among these include the International Tropical Timber Organization's (ITTO), *Sustainable Forest Management* initiative, and the United Nations' *Reducing Emission from Deforestation and Degradation Plus* (*REDD+*) initiatives. The latter includes forest conservation, sustainable forest management, and the carbon that can be stored by 'reforestation', including through industrial tree plantations. The European Union's *Forest Law Enforcement, Governance, and Trade (FLEG-T)* program, which aims to minimize illegal logging by boosting sustainable management of the forest resource, improving governance, and increasing the trade of legally procured wood is another successful example. Funding mechanisms are available to support these initiatives, and in the case of the FLEG-T, the *Voluntary Partnership Agreement* (*VPA*) scheme has been initiated between the EU and several tropical countries unilaterally, to facilitate sustainable wood trade (Saunders 2014).

Over the last decade, companies are increasingly assessed in terms of their compliance to the *Environmental, Social, and Governance (ESG)* indicators. ESG refers to the three most important criteria which determines the social impact and long-term viability of a business enterprise or organization. These criteria will shape the overall

Fig. 11.3 Common Timber Certification schemes

financial performance (i.e., return and risk) of enterprises in the future. In fact, the ESG scores are being increasingly used as an evaluation tool by asset managers and financial institutions to rate and compare enterprises' performance on the ESG perspective (Ratnasingam 2020).

Environmental criteria define how a company behaves as a steward of the environment. The way it engages and connects with employees, suppliers, consumers, and the community in which it operates is assessed using social criteria. Governance deals with the leadership of a company, as well as remuneration scheme, audits, internal controls, and shareholder rights.

The ESG indicators are becoming key points to be considered by potential investors in enterprises and companies, and are recognized as market requirements as: (1) environmental, social, and governance (ESG) criteria are becoming more widely used by traders to evaluate organizations in which to invest, (2) many mutual funds, brokerage, and robo-advisors now provide products that that adopt ESG criteria, and (3) ESG criteria also assist traders avoid enterprises and companies that may pose extra financial risk due to their environmental and related activities.

Recently, the *Sustainability Policy Transparency Toolkit* (abbreviated as *SPOTT*) has emerged as another tool to assess commodity producers, processors, and traders on their public disclosure concerning their rules and practices associated with environmental, social, and governance (ESG) matters (Anghela et al. 2019). SPOTT ratings are available for tropical forestry, palm oil, and natural rubber enterprises, on an annual basis, which tracks and benchmarks their progress through the years. Investors, shareholders, buyers, and other stakeholders apply the SPOTT tests to inform and manage ESG threats and increase transparency across industries, apart from incentivizing the implementation of corporate best practices (Ratnasingam 2020).

As environmental consciousness grows around the world, against the threat of irreversible global climate change, consumers are insisting that sustainable and green practices become a norm in the furniture industry (Kim et al. 2010; Liu et al. 2015). On this ground, the application of recycled fibers is a norm in the paper and paperboard industries, particularly for packaging purposes. Almost all paper products meant for packaging of furniture are manufactured using a large proportion of recycled fibers, in line with the market requirement (Fig. 11.4).

Packaging can play several roles in a circular economy (Anghela et al. 2019). Among the roles are: (i) packaging can contribute to reducing waste (packages have a longer use life), (ii) packaging can be reused (B2B and B2C trade interchanging their packaging), (iii) packaging materials may be recycled, and (iv) packaging materials with high energy content, such as fiber and plastic, which cannot be reused or recycled, may serve as fuel in energy generation.

Packaging is, in fact, a key component of the circular economy. Reverse logistics allow waste materials to be explored as raw material for new products and industries. As a circular economy becomes mainstream throughout the world, green options related to energy generation will be preferred and demanded by consumers worldwide (Maraseni et al. 2017).

Fig. 11.4 Recycled carton box

Other green initiatives observed in the furniture industry are the increasing use of water-based coatings or finishes and natural adhesives (i.e., soy-based, tannin-based adhesives), which is a marked shift from the VOCs and other toxic substances emitting coatings and adhesives (Morris and Dunne 2004).

The use of plastics mainly for packaging, considered an environmental pollutant, is also being reduced appreciably in the course of time in many fields, and a comparable trend is likewise obvious in the furnishing industry (Xu et al. 2020).

Over the last few decades, customers throughout the world are becoming increasingly green and environment conscious and demanding that furniture must have a *low carbon footprint*. In this context, the global furniture industry is expected to embrace green and clean manufacturing concepts extensively (Iritani et al. 2015). As the global industry evolves towards a low-carbon economy, information on the 'cradle to grave', or rather environmental effect based on the *life-cycle assessment* (LCA) of furniture, will be requested before furniture is purchased in the future.

Summary

- Green and clean manufacturing practices will transform the furniture industry, which is presently, stigmatized as being the dangerous, dirty, and difficult (3D) industry.
- Standardization, environmental compliance, and certification are effective marketing tools for the furniture industry.
- Occupational safety and health of workers must be in accordance with the SDGs.

References

Anghela MD, Bravi L, Murmura F (2019) Wood furniture SMEs approaches towards circular economy—a literature review. In: Proceedings of the 1st conference on quality innovation and sustainability, Valencia, Portugal. 6–7 June 2019, pp 54–60

Böhm M, Salem MZ, Srba J (2012) Formaldehyde emission monitoring from a variety of solid wood, plywood, blockboard and flooring products manufactured for building and furnishing materials. J Hazard Mater 221:68–79

Bulian F, Fragassa C (2016) VOC emissions from wood products and furniture: a survey about legislations, standards and measures referred to different materials. FME Trans 44:358–364

Chen J, Innes JL, Kozak RA (2011) An exploratory assessment of the attitudes of Chinese wood products manufacturers towards forest certification. J Environ Manage 92(11):2984–2992

Cinar H, Erdogdu M (2018) Eco-design: effects of thickness and time in service for wood-based boards on formaldehyde emission. Forest Prod J 68(4):405–413

Daian G, Ozarska B (2009) Wood waste management practices and strategies to increase sustainability standards in the Australian wooden furniture manufacturing sector. J Clean Prod 17(17):1594–1602

Espinoza O, Buehlmann U, Smith B (2012) Forest certification and green building standards: overview and use in the US hardwood industry. J Clean Prod 33:30–41

Fet AM, Skaar C (2006) Eco-labeling, product category rules and certification procedures based on ISO14025 Requirements. Int J Life Cycle Ass 11:49–54

Holopainen J, Toppinen A, Perttula S (2015) Impact of European Union timber regulation on forest certification strategies in the finnish wood industry value chain. Forests 6(8):2879–2896

Iritani DR, Silva DAL, Saavedra YMB et al (2015) Sustainable strategic analysis through Life Cycle Assessment (LCA): a case study in a furniture industry. J Clean Prod 96:308–318

Kim S, Choi YK, Park KW et al (2010) Test methods and reduction of organic pollutant compound emissions from wood-based building and furniture materials. Bioresource Technol 101(16):6562–6568

Liu X, Mason MA, Guo Z et al (2015) Source emission and model evaluation of formaldehyde from composite XE "Composites" and solid wood furniture in a full-scale chamber. Atmos Environ 122:561–568

Maraseni TN, Son HL, Cockfield G et al (2017) The financial benefits of forest certification: case studies of Acacia growers and a furniture company in Central Vietnam. Land Use Policy 69:56–63

Mirka GA (2005) Development of an ergonomics guideline for the furniture manufacturing industry. Appl Ergon 36(2):241–247

Morris M, Dunne N (2004) Driving environmental certification: its impact on the furniture and timber products value chain in South Africa. Geoforum 35(2):251–266

Nizialek I, Podobas I, Solka M et al (2018) The application of TQM in the wood sector company. Ann Warsaw Univ Life Sci Forest Wood Technol (102): 97–102

Ratnasingam J (2020). Environmental compliance in the furniture industry—a review. Tech. Note, no 11. IFRG Publication, Singapore

Ratnasingam J, Macpherson TH, Ioras F (2008a) An assessment of Malaysian wooden furniture manufacturers' readiness to embrace chain of custody (COC) certification. Holz Roh Werkst 66(5): 339–343

Ratnasingam J, Macpherson TH, Ioras F et al (2008b) Chain of custody certification among Malaysian wooden furniture manufacturers: status and challenges. Int For Rev 10(1): 23–28

Ratnasingam J, Yoon CY, Ioras F (2013) The effects of ISO 9001 quality management system on innovation and management capacities in the Malaysian furniture sector. Bulletin of the Transilvania University of Brasov. Forestry, Wood Industry, Agricultural Food Engineering. Series II, vol 6, no 1, p 63

Ratnasingam J, Scholz F (2015) Dust emission characteristics in the bamboo and rattan furniture manufacturing industries. Eur J Wood Wood Prod 73(4):561–562

Ratnasingam J, Ioras F, Abrudan I (2012) An evaluation of occupational accidents in the wooden furniture industry—a regional study in Southeast Asia. Saf Sci 50:1190–1195

Ratnasingam J, Lim CL, Ab Latib H (2019) A Comparison of the abrasive sanding dust emission of oil palm wood and Rubberwood. BioResources 14(1):1708–1717

Ratnasingam J, Ramasamy G, Ioras F et al (2016) Assessment of dust emission and working conditions in the bamboo and wooden furniture industries in Malaysia. BioResources 11(1):1189–1201

Ruddell S, Stevens JA (1998) The adoption of ISO 9000, ISO 14001, and the demand for certified wood products in the business and institutional furniture industry. Forest Prod J 48(3):19

Stevens J, Ahmad M, Ruddell S (1998) Forest products certification: a survey of manufacturers. Forest Prod J 48(6):43–46

Saunders J (2014) Certified products and EUTR compliance in the furniture sector. Chatham House-EER PP 2014/09, London, UK

Tong R, Zhang L, Yang X et al (2019) Emission characteristics and probabilistic health risk of volatile organic compounds from solvents in wooden furniture manufacturing. J Clean Prod 208:1096–1108

Xiong X, Ma Q, Yuan Y et al (2020) Current situation and key manufacturing considerations of green furniture in China: A review. J Clean Prod 267: 121957

Xu X, Hua Y, Wang S et al (2020) Determinants of consumer's intention to purchase authentic green furniture. Resour Conserv Recy 156: 104721

Chapter 12
Strength Design and Furniture Testing

Furniture is an engineering structure and is often subjected to a variety of loads during its service life. Therefore, the design and construction of furniture must incorporate the necessary strength features, to ensure it meets the strength and durability requirements in accordance with the standard of the target market. Failure in furniture is often attributed to its joints, which is the weakest link in the structure. Component failure rarely occurs in furniture, though the component size is determined by the joint design and dimension. Furniture testing is often conducted to assess the strength and durability of furniture before it is approved for use.

12.1 Introduction

Product engineering is defined as the development of a product primarily based on scientific concepts and strategies, under the influence of physical and mechanical criteria (Anon 2014). In this context, furniture design and development are often considered as an artwork. At the same time, the engineering aspects related to the structural strength and durability is considered a science. Although aesthetic appeal of furniture is deemed most important, the structural strength and durability which affects its functionality in service is of equal importance (Prekrat et al. 2011).

In essence, a methodical design technique needs to be followed to produce structurally sound, secure, and durable furniture. The principles of structural mechanics are applicable and may be implemented to furniture, which are considered complex structures. Furniture frames are statically indeterminate because the members are shaped, and with inconsistent cross sections (Erdil et al. 2004). Further, furniture joints are elastic in nature. The higher the modulus of elasticity (MOE), the less it deforms, and vice versa. As a result, furniture is indeed an engineering structure with semi-rigid joints (Ratnasingam and Ioras 2013, 2015).

J. Ratnasingam, *Furniture Manufacturing*, Design Science and Innovation,
https://doi.org/10.1007/978-981-16-9412-7_12

12.2 Performance Testing for Furniture Engineering

The engineering of furniture often involves performance testing, as a means of quality assurance (Smardzewski 1998; Eckelman 2003). Performance tests are simply an accelerated tests that are performed to evaluate if the furniture can fulfill its intended function. Such tests will identify weak points to improve the overall structural strength and durability. Consequently, the main objectives of the furnishings overall performance tests are to: (1) *provide quantitative feedback to the designers to improve the structure,* (2) *identify weak spots in the design that may have been overlooked, but may arise during service, and* (3) *affirm that the furniture is fit for use* (Ho and Eckelman 1994).

Before developing any furniture performance tests, information on its condition of use, estimates of the loading pattern and its frequency should be gathered. Based on this information, an appropriate test protocol can be devised that simulates the conditions in service. It must be appreciated that furniture is subjected to regular loads, and the occasional abusive loadings, at some point during its service life. In this context, the performance tests must take these events into consideration (Eckelman 1978).

When the furniture is new and retains a high degree of its structural integrity, it can withstand a wide range of loading patterns, even abusive loads on occasion. As the furniture's strength deteriorates over time, a point is reached when the load exceeds the residual strength, and the furniture fails (Eckelman 1978). The 'first-crossing' concept of failure, which is essentially based on the cumulative damage theory as outlined in structural mechanics, describes the point at which the furniture structure collapses. In essence, it is hypothesized that whenever furniture is subjected to a load, it is scarcely damaged and hence hardly weakened as a result of this motion. New furniture with a high initial strength can withstand both common and 'abusive' unusual loads. The fatigue energy (or fatigue resistance) of the components and joints reduces over time and use, and finally a point is reached where the applied

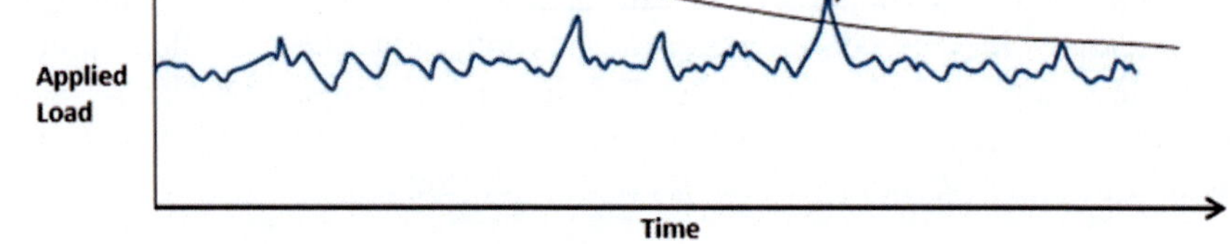

Fig. 12.1 The first crossing concept in furniture structure (Adapted from Eckelman 1978)

load, sometimes an abusive load, exceeds the residual resistance of the furniture structure, causing failure (Fig. 12.1).

Based on structural mechanics, furniture failure is often due to fatigue loads, rather than static or impact loads (Ratnasingam et al. 1997). As a result, furniture performance testing incorporates the first-crossing concept by adopting the 'cyclic stepped loading' method which is a fatigue-oriented test method, that introduces loads in installments to the structure until failure occurs (Hearn 1997; Eckelman 2003).

Traditional furniture performance testing is not only time consuming, but expensive too. It starts with making the necessary prototypes, which is subjected to repeated testing until approval is obtained. Such tasks may be unacceptable for many export-oriented furniture manufacturers who work on fast turn-around of business (Anon 2014; Ratnasingam 2019). Inevitably, the need for less time-consuming furniture performance testing is highly desirable.

12.3 Advanced Tools for Furniture Engineering

The analysis of furniture structure is complicated, as furniture structures are made up of many parts and connections, which deform non-linearly under load (Durka et al. 1996). Similarly, the geometric shape of furniture cannot be presented linearly, and therefore 3-dimensional illustration is constantly essential. With the advent of technology, this task has been simplified. The application of computer systems and analytical software programs has rendered the analysis of furniture structure easier, and consequently, calculations and the need for physical prototype is not a necessity (Hasan 2019).

Finite element methods (FEM) are the most common analytical tool used to evaluate the strength design of furniture frames, and the resultant stiffness matrix structural analyses will pin-point the weakness points in the structure (Chandrupatla and Belegundu 1991; Smardzewski 2015). In essence, developments in computer technology and programs have enabled the structural integrity of the furniture frames to be accurately done in a short period of time, without the need for prototypes. The multiple rigidity/strength analysis that can be carried out quickly comes in very useful, as design and structural alterations can be achieved in a short period of time (Fig. 12.2). This is indeed a major accomplishment in product engineering because it allows structural weaknesses in furniture design to be identified quickly, using 3D prototype models in computers, instead of a physical prototype.

The earliest work in this field of product engineering was in the late 1960s, using FORTRAN computer language-based algorithms to perform such analysis (Eckelman 2003). The load-deformation characteristics of the joints, determined by its stiffness and tension are essential parameters in such evaluations, using the ANSYS Parametric Design Language (APDL) to construct the analytical models. This improvement allowed the level of reliability, design stress values and structural

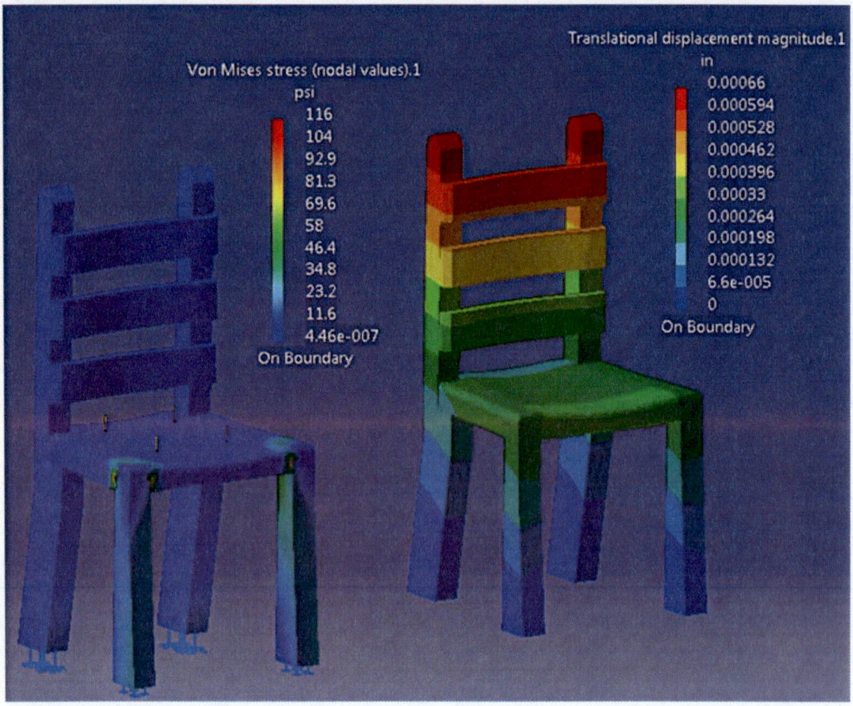

Fig. 12.2 Finite element analysis of a chair

modeling to be integrated into the computations through finite detail strategies (FEM) and joint design.

12.4 Principles of Strength Design of Furniture

Generally, furniture designers tend to pay greater attention to aesthetic appeal of the piece, rather than the engineering aspects of the structure (Smardzewski 2015). This is simply due to the fact that most designers lack knowledge on the optimum sizes of material to be used as members, apart from its design safety. Therefore, it is common to find furniture with over-sized components and joints, attributed to the lack of application data on furniture structures as well as the limited use of structural mechanics in designing furniture (Hu et al. 2019).

On the other hand, customers are also too engrossed with the aesthetic appeal of furniture, its price, and functionality, rather than the science, engineering, and technology that goes into its design (Ratnasingam 2019). Unfortunately, such an approach is inappropriate for a fashion item such as furniture. Normally, furniture manufacturers can produce similar furniture design of comparable price-points, using two different wood species, such as Walnut and Pine, but the structural integrity of the two designs will be markedly different. The point is, it is rather misleading to judge a piece of furniture based on its price-tag, which may not be a good representation of its quality let alone its structural durability and functionality (Ratnasingam 2019).

Table 12.1 shows the properties of a few selected wood species, aimed at highlighting the variable nature of wood. However, any wood species can be used as a furniture material, the only point to be considered is the size of the material to be used to support the load it is expected to sustain (Zhang et al. 2001, 2003; Thompson et al. 2002). Therefore, a wood species with lower modulus of elasticity (MOE) and modulus of rupture (MOR) will necessarily require members of larger cross section to support the given load, compared to a wood species with higher MOE and MOR. Likewise, furniture joints, with higher MOE will deform less, and vice versa, in accordance with the concepts of structural mechanics. The elasticity of furniture joints renders such a characteristic (Hu et al. 2019).

Furniture engineering relies upon an in-depth knowledge of loading pattern, physics, and structural mechanics to analyze and predict how the structure supports its own weight as well as the imposed loads (Eckelman 2003). The criteria which determine a good furniture structure is both serviceability, or rather functionality, and its strength to support the design loads. Inevitably, the furniture structure must be both strong and stiff to withstand the service loads (Bao et al. 1996; Erdil et al. 2004). The load applied on to a furniture structure is distributed and transmitted through the many structural elements and joints (Fig. 12.3). These forces can manifest themselves as tension, compression, shear, and bending, or even bending moment (Ratnasingam and Ioras 2010, 2011a, b; Šimek and Sebera 2015).

On this principle, the strength of furniture is dependent on the strength of the component members and the joints. Therefore, the furniture designer must appreciate the properties of wood, and capitalize on it in furniture design (Durka et al. 1996; Ratnasingam and Muttiah 2017). Before going any further, some important strength properties of wood should be explained (John 1992).

(i) *Modulus of Elasticity* (MOE) reflects the resistance of the material to deform elastically when under stress. In other words, materials with higher MOE are stiffer, and vice versa.

Table 12.1 Comparative properties of some common timbers

	Lignum vitae	Balsa (*Ochroma pyramidale*)	Kapur (*Dryobalanops spp.*)	Meranti (*Shorea spp.*)
Density at 12% moisture content	*1260*	*160*	*640*	*460*
Modulus of Elasticity (N/mm^2)	*14,900*	*3,400*	*13,000*	*12,200*
Modulus of Rupture (N/mm^2)	*127.9*	*21.3*	*126*	*87.6*
Compression // to grain (N/mm^2)	*78.6*	*14.9*	*65.3*	*42.1*
Shear Strength (N/mm^2)	*22.6*	*2.1*	*13.7*	*10.1*
Janka Hardness (N)	*20,000*	*400*	*4400*	*3100*

Source Anon (1989)

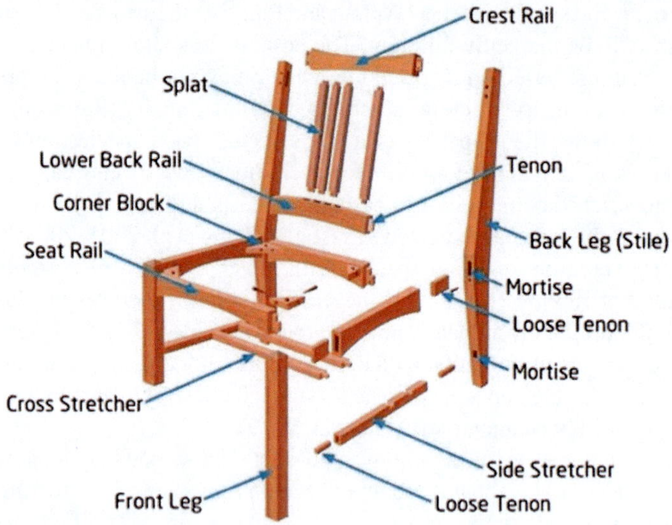

Fig. 12.3 Structure of a chair—its members and connections

(ii) *Modulus of Rupture* (MOR), also known as *flexural strength*, is a highest stress point before the material fails, in a flexure test.
(iii) *Crushing Strength* also called *compressive strength parallel to the grain* is the resistance of the material under a compressive load.
(iv) *Shear Strength* is the resistance of the material to sliding failure along the plane, parallel to the direction of force application.
(v) *Janka Hardness* is a measure of a wood's resistance to denting and wear.
(vi) *Bending Moment* is a product of force and distance, or a lever arm, which produces a turning effect or torque.

The above strength properties dictate the movements of members and joints in the furniture structure when a load is applied. Load (Case et al. 1983). Against this background, the selection of materials for application in furniture must be based on its density, modulus of elasticity, modulus of rupture, and compressive strength (Table 12.2).

The density of a material is simply its mass over a unit volume, and it affects its ease of handling and workability. MOE, MOR, and compressive strength affects the material's ability to withstand load. Other properties including durability, flammability, workability, market availability, environment-friendly, and its cost are often taken into consideration when choosing materials for furniture.

The *stiffness of the furniture structure* is dependent on the counter-acting forces in equilibrium from the multiple components in the structure. The many components in a furniture structure are connected through joints, which also happens to be the weakest link in the complete structure. Consequently, furniture often fails at the joints.

Table 12.2 Comparative strength properties of materials

Material	Density (kg/m^3)	Modulus of elasticity (N/mm^2)	Modulus of rupture (N/mm^2)	Compressive strength (N/mm^2)
Wood (*Shorea* spp.)	640	10,500	54	27
Bamboo (*Dendrocalamus* spp.)	620	20,000	180	80
Rattan (*Calamus* spp.)	530	3,000	70	29
Metal (*Steel*)	8050	210,000	140	140
Plastic (*PVC*)	380	11,000	25	54
Medium Density Fiberboard (*MDF*)	600	2,700	24	10

Source Anon (1989)

Figure 12.4 shows the typical failure mode in a chair structure. The four vertical legs in the chairs are connected to one another through side-rails and stretchers. The main purpose of the stretcher is to provide extra bracing to the legs, minimizing horizontal movement relative to each other. The structure is strengthened further by the cross rails. These components react keeping the counter forces in the structure, in equilibrium (Smardzewski 1998).

Figure 12.4 also reflects the distribution of the load, which encompasses both, the own weight and applied load, within the structure of the chair. Previous research on furniture engineering has emphasized the need to: (1) use the proper connections, i.e., joints and fittings, based on the loading pattern and the intended use of the furniture, (2) ensure the forces in the structure are in equilibrium, and (3) establish the allowable design stress for the furniture to prevent premature (Eckelman 1978). Following these guidelines will pave the way for successful furniture engineering,

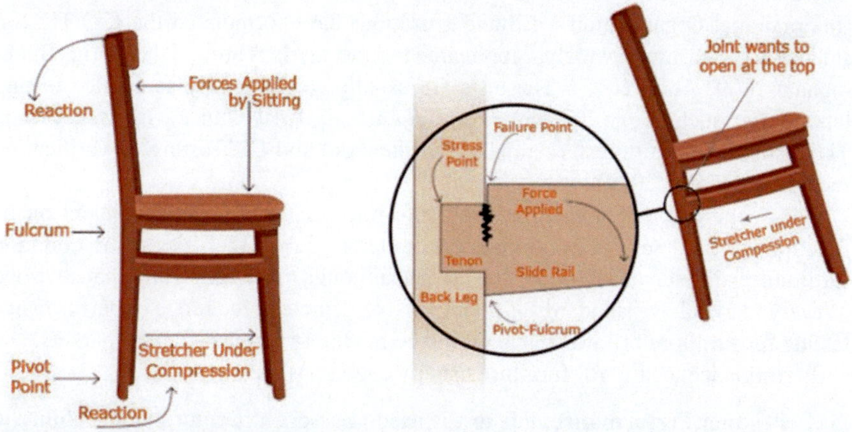

Fig. 12.4 Load diagram in a chair

although the aspects related to the aesthetic appeal, comfort, and ergonomics must also be kept in mind to ensure the furniture is accepted by the customer.

12.5 Types of Furniture Testing

Product reliability and performance testing of furniture are standard practices in the global furniture industry, although information and computer technology (ICT) has revolutionized product development processes. Such tests are mandated in almost all countries throughout the world to ensure product durability and safety. Increasingly products are being tested under actual or simulated service conditions (Anon 2014). The underlying reason for carrying out performance testing is to identify the circumstances under which the product has a risk of failure, and the consequences of such failures. Generally, the first prototypes would have many design and engineering flaws, which are rectified based on the performance test data. In fact, it has been shown that the reliability of a new product design might be as low as 15–50% of the final product. In this respect, as highlighted by Ratnasingam (2019), prototype testing remains one of the foremost requirements that must be met by manufacturers to meet the customer's expectations. Though it is a time-consuming procedure, it's benefits outweighs the cost involved, and would eventually contribute towards customer satisfaction.

Furniture testing standards are aplenty, and among the common ones are the DIN (*Deutsches Institut für Normung*), BS (British Standard), and BIFMA (Business and Institution Furniture Manufacturers Association), Standard Australia, ASTM (American Society for Testing and Materials), JIS (Japanese Industrial Standard). The standards are market requirements for specific jurisdictions or countries and may be cumbersome for manufacturers to work with many different standards. To harmonize these standards, the CEN (European Committee for Standardization) and ISO (International Organization for Standardization) have formulated the ISO/TC 136 standards for furniture, which is applicable to most markets around the world (Buckingham 2009; Anon 2015). The tests are usually conducted by accredited testing laboratories such as the SGS *(Société Générale de Surveillance)*, Intertek Group, TÜV Rheinland, Furnitest Testing & Certification, and UL Testing & Verification Inc. (Ratnasingam 2019).

The scope of performance testing conformity of furniture is dependent on its functionality and service conditions encountered. Table 12.3 lists some common furniture testing standards used in the market, although the list is not an exhaustive one as every market have standards of their own. Nevertheless, the scope of performance testing for furniture follows a general theme in almost all the different markets.

Performance testing for furniture usually encompasses:

(1) **Product Performance**: this test is used to assess a furniture item's ability to endure wear and tear during normal, expected use (Anon 2014). Performance

Table 12.3 List of some selected furniture standards

Selected furniture standards from different jurisdictions
• EN 527 Office furniture—Worktable and desk • EN 1335 Office furniture—Office work chair • ANSI/BIFMA X 5.1 Office Seating • DIN 4551 Office furniture; revolving office chair with adjustable back with or without arm rests, adjustable in height • EN 581 Outdoor furniture—Seating and tables for camping, domestic, and contract use • EN 1728:2014 Furniture—Seating—Test methods for the determination of strength and durability—updated in 2014 • EN 1730:2012 Furniture—Test methods for the determination of stability, strength, and durability • BS 4875 Furniture. Strength and stability of furniture. Methods for determination of stability of non-domestic storage furniture (British Standard) • EN 747 Furniture—Bunk beds and high beds—Test methods for the determination of stability, strength, and durability • EN 13150 Workbenches for laboratories—Safety requirements and test methods • EN 1729 Educational furniture, chairs, and tables for educational institutions • RAL-GZ 430 Furniture standard from Germany • NEN 1812 Furniture standard from the Netherlands • GB 28007-2011 Children's furniture – General technical requirements for children's furniture designed and manufactured for children between 3 and 14 years old • BS 5852: 2006 Methods of test for assessment of the ignitability of upholstered seating by smoldering and flaming ignition sources • BS 7176: Specification for resistance to ignition of upholstered furniture for non-domestic seating by testing composites
ISO/TC 136 Furniture Standards
• ISO 3055:1985: Kitchen equipment—Coordinating sizes • ISO 4211-2:2013: Furniture—Tests for surface finishes—Part 2: Assessment of resistance to wet heat • ISO 4122-3:2013: Furniture—Tests for surface finishes—Part 3: Assessment of resistance to dry heat • ISO 4211-4:1988: Furniture—Tests for surfaces—Part 4: Assessment of resistance to impact • ISO/DIS 4211-5: Furniture—Tests for surface finishes—Part 5: Assessment of resistance to abrasion • ISO 4211:1979: Furniture—Assessment of surface resistance to cold liquids • ISO/AWI 4769: Hardware for furniture—Strength and durability of hinges and their components—Hinges pivoting on a vertical axis • ISO 7170:2005: Furniture—Storage units—Determination of strength and durability • ISO 7171:2019: Furniture—Storage units—Test methods for the determination of stability • ISO 7172:2019: Furniture—Tables—Determination of stability • ISO 7173: 1989: Furniture—Chairs and stools—Determination of strength and durability • ISO/WD 7173: Furniture—Chairs and stools—Determination of strength and durability • ISO 7174-1:1988: Furniture—Chairs—Determination of stability—Part 1: Upright chairs and stools • ISO 7174-2:1992: Furniture—Chairs—Determination of stability—Part 2: Chairs with tilting or reclining mechanisms when fully reclined and rocking chairs

(continued)

Table 12.3 (continued)

Selected furniture standards from different jurisdictions
• ISO 7175-1:2019: Furniture—Children's cots and folding cots for domestic use—Part 1: Safety requirements • ISO 7175-2:2019: Furniture—Children's cots and folding cots for domestic use—Part 2: Test methods • ISO 8191-1:1987: Furniture—Assessment of the ignitability of upholstered furniture—Part 1: Ignition source: smouldering cigarette • ISO 8191-2:1988: Furniture—Assessment of ignitability of upholstered furniture—Part 2: Ignition source: match-flame equivalent • ISO 9098-1:1994: Bunk beds for domestic use—Safety requirements and tests—Part 1: Safety requirements • ISO 9098-2:1994: Bunk beds for domestic use—Safety requirements and tests—Part 2: Test methods • ISO 9221-1:2015: Furniture—Children's high-chair—Part 1: Safety requirements • ISO 9221-2:2015: Furniture—Children's high-chair—Part 2: Test methods • ISO 10131-1:1997: Foldaway beds—Safety requirements and tests—Part 1: Safety requirements • ISO 10131-2:1997: Foldaway beds—Safety requirements and tests—Part 2: Test methods • ISO/WD 19682: Furniture—Tables—Test methods for the determination of stability, strength, and durability • ISO 19833:2018: Furniture—Beds—Test methods for the determination of stability, strength, and durability • ISO 21015:2007: Office furniture—Office work chairs—Test methods for the determination of stability, strength, and durability • IS0 21016:2007: Office furniture—Tables and desks—Test methods for the determination of stability, strength, and durability • ISO/DIS 23767: Children's furniture—Mattresses for cots and cribs—Safety requirements and test methods • ISO/DIS 23769: Furniture—Mattresses—Test methods for the determination of functional characteristics • ISO 24496:2017: Office furniture—Office chairs—Methods for the determination of dimensions

Source Malaysian Furniture Council

testing is typically a key strategy of a manufacturer's quality management strategy, and usually involves:

Surface testing—Determines a furniture item's resilience to water, cleaning agents, as well as scratches and abrasion.

Rapid ageing testing—Rapid ageing tests assess a product's resilience to the effects of moderate or hot over time.

Packaging suitability—Drop and vibration testing are used to test the packaging appropriateness for furniture.

(2) **Mechanical Testing**: this test aims to assess the mechanical or strength performance of furniture (Ratnasingam 2019). Although not a regulatory requirement, mechanical testing is often undertaken by furniture manufacturers as a

guarantee of product safety. Mechanical testing can be done in a variety of ways, including:

Structural testing—Structural testing evaluates the product's static and dynamic load-carrying capacity, deflection characteristics, swivel length, and structural stability (Fig. 12.5)

Component testing—examines the strength and durability of product components, and hardware, including chair frames, hinges, door locks, and drawer guides/rollers (Fig. 12.6).

Mechanical safety—Mechanical safety testing is meant to examine hazards that may arise from sharp points and edges of the furniture, which may injure the users.

(3) **Electrical Safety Testing**: is important for many modern furniture which integrate electrical or electronic components to provide illumination or to make it easier to operate or manage electrical features/appliances. Inevitably, such

Fig. 12.5 Furniture testing equipment—mechanical tests (Courtesy of the Malaysian Furniture Council)

Fig. 12.6 Furniture component/hardware testing equipment (Courtesy of the Malaysian Furniture Council)

furniture is mandated to undergo electrical safety testing in many markets (Anon. 2014). The purpose of electrical safety testing in such furniture is to eliminate any electrical hazards that may endanger the safety of its users, or even eliminate the possibility of fire and electrical shock.

(4) **Flammability Testing**: furniture is frequently made up of flammable materials, such as wood, upholstery textiles, adhesives, coatings, and varnishes (Erdil et al. 2004). Hence furniture may serve as an indoor fire ignition point, and for this reason, necessitates flammability testing, which is a mandatory requirement in most markets. Flammability of a material is evaluated by its resistance to heat, flame, and its burning characteristics as elaborated below:

Ignition testing—Ignition testing determines furniture's resistance to ignition when exposed to excessive heat or an open flame and is measured by the elapsed time between initial exposure and ignition (Fig. 12.7).
Flame spread testing—Flame spread testing evaluates the rate at which the fire spreads from the point of ignition.
Heat-release characterization—The heat released from burning furniture can adversely affect ambient temperature conditions in an indoor fire, causing a flashover. It indicates the quantity of heat build-up contributed by the burning furniture in a concealed area.

(5) **Chemical Emissions and Chemical Content**: in most jurisdictions across the world, the use of chemicals in furniture and the regulation of chemical emissions from furniture are strictly regulated. In the European Union (EU), the REACH (Registration, assessment, Authorization, and restrict of chemicals) regulation is applicable to most chemical containing products offered within the EU market. Within the U.S., the Consumer Product Safety Improvement Act (CPSIA) bans or substantially restricts the use of specific chemicals in products for children (Anon. 2014).

Types of chemical testing undertaken for furniture include:

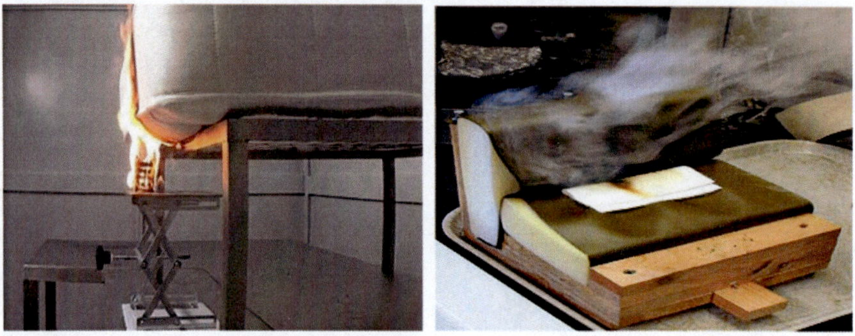

Fig. 12.7 Furniture flammability testing (Courtesy of the Malaysian Furniture Council)

Chemical content testing—Chemical and microbiological content testing is frequently used to evaluate the presence of heavy metals, consisting of lead (Pb), biocidal substances, phthalates, flame retardants, and materials whose use may be restricted or banned.

Chemical emissions testing—Furniture may emit VOCs and formaldehyde over time, resulting in poor indoor air quality. To eliminate such emissions, product certifications for environmental compliance should be undertaken.

(6) **Environmental Sustainability**: product specifiers, procurement specialists, and buyers are increasingly concerned about 'green' furniture, i.e., furniture made with more environmentally friendly materials, or manufactured using techniques and processes that have a lower environmental impact. For instance, the LEED Rating System for green buildings allows extra qualifying points when the green furniture is used in the building (Anon 2014). Furthermore, evidence of environmental sustainability can address consumer concerns and help manufacturers differentiate their furniture in a crowded marketplace.

12.6 Statistical Quality Control

Since the last few decades, one of the most prominent tools for quality assurance that has been adopted throughout the global furniture manufacturing industry is the *Statistical Quality Control (SQC)*. SQC is a method that applies statistical analyses throughout the manufacturing operations to enhance the quality of (Simanová and Gejdoš 2015). Statistical records are used to closely monitor and control repetitive production operations. In addition, SQC controls the quality characteristics of the product manufactured, by managing the work methods and machines employed to the benefit of the company as well as the workers.

The basic tool for SQC is the quality control charts. The primary goal of the quality control charts is to identify and reduce variation. The sources of variance must be identified and minimized, if not eliminated, when the process is out of control. Raw materials, equipment, and operating procedures/processes could all contribute to variations. Only random variations are tolerated in manufacturing processes. Figure 12.8 shows an example of a control chart that can be used to identify product quality variations.

Control limits (CL) refers to the tolerable quality limits for the product in the factory, while *specification limits* (SL) refer to the acceptable quality of the product by the customers. The population mean (χ) and standard deviation (σ) determine these limits, and the difference between the upper and lower limits is usually 3σ.

Another tool that is incorporated into the SQC initiatives by companies is the Pareto analysis. This analysis is based on the premise that approximately 80% of problems are caused by 20% of the possible causes. Consequently, SQC and Pareto evaluation can help an organization prioritize major quality problems to deal with and focus on the vital few problems, rather than the inconsequential many.

Fig. 12.8 Distribution of variation limits for a product

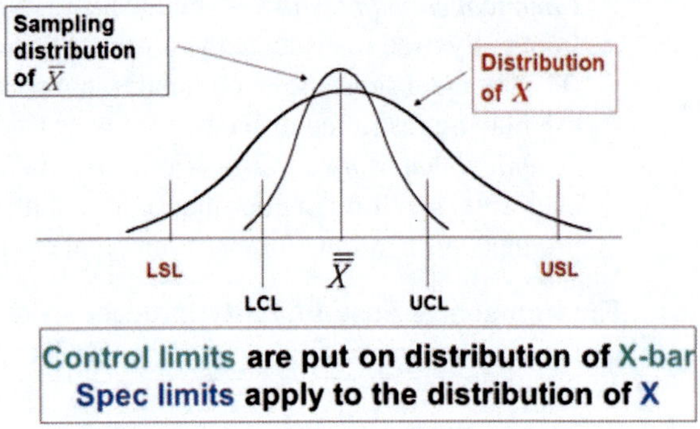

Several actions can be taken to ensure that the Pareto analysis is effective as an SQC initiative. First, a standardized listing of a clear set of criteria for non-conformities must be done (Vickery et al. 1997). Second, the non-conformities should be labeled consistently. For instance, in furniture manufacturing, one worker may identify a defect as *torn grain*, while another may call it *fuzzy grain*. Although these defects may appear similar, but their causes may be different. In this context, clear identification of non-conformities will reduce and eliminate confusions on the factory shopfloor. One practical way to achieve this is to display all non-conformities, defects, and blemishes, identified and labeled accurately in a manner visible to everyone in the organization.

Pareto analysis requires the frequency of every non-conformity to be recorded and charted, from the lowest to the highest frequency of incidence. Figure 12.9 illustrates how the typical Pareto chart is displayed. Apart from creating a chart based on continuous readings from the factory shopfloor, Pareto charts could also

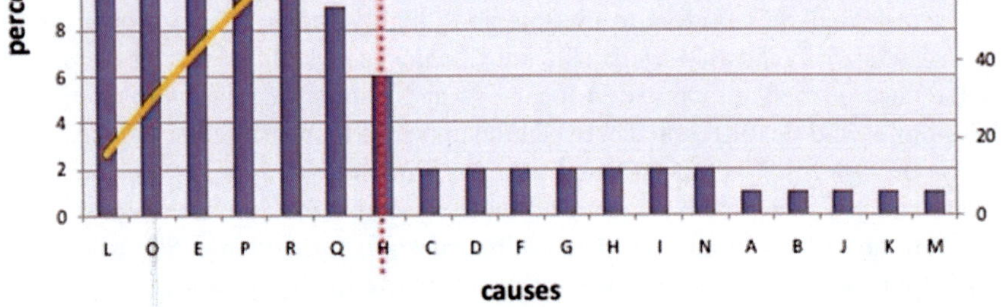

Fig. 12.9 Pareto analysis

be built based on measurements, such as good or bad, accept or reject, go/no-cross, or bypass or fail criteria (Vickery et al. 1997). What is required is the frequency recordings of the non-conformity incidences from the lowest to the highest.

Several continuous improvement principles should be followed in order for SQC techniques to be implemented effectively. SQC begins by implementing continuous improvement ideas by taking measurements at all stages of the production process. This allows faults to be spotted early on, which saves money when compared to discovering a flaw after the entire production has been completed. In addition, for SQC to be effective, furniture manufacturing companies should involve all levels of employees (i.e., operators, operator assistants, supervisors, and senior management) in the decision-making process. This is particularly important in the manufacturing industry predominated by small and medium companies (SMEs), frequently family controlled businesses (Ratnasingam 2019). In such circumstances, there is usually substantial experience and keen eyes among the more experienced and skilled employees, for spotting production ineffectiveness that must be taken advantage of.

Such a transformation in quality control approach is highly desirable in the furniture manufacturing industry, as it shifts towards embracing production engineering principles, where quality is measured quantitatively, instead of qualitatively. Adopting SQC system in furniture manufacturing based on continuous improvement practices will not only improve overall product quality, but will also improve customer satisfaction, business competitiveness, and also reduce manufacturing cost (Simanová and Gejdoš 2015).

12.7 Recommendations for Furniture Manufacturers

In the global furniture markets, for manufacturers to fulfill customer expectations and market requirements, it is important for them to (1) appreciate regulatory requirements and market needs, (2) carry out necessary performance testing early in the product development phase, and (3) seek out and improve design and structural flaws of the furniture based on the test results (Ratnasingam 2019). Such an approach is important for furniture, which is regarded as a fashion and merchandize, which must be designed and manufactured to offer the highest quality, customer satisfaction, and value for money.

As the world moves into the era of Industry 4.0, the furniture manufacturing industry will also have to tag along and embrace this concept. The implementation of key enabling technologies (KET) of Industry 4.0, notably 3D modeling and printing, computer-aided design (CAD), computer numerical control (CNC) machining centers, and picture visualization, is becoming the norm in furniture manufacturing, as the industry embraces digitalization. These technologies will shorten the design, engineering, and product development phases in furniture manufacturing, apart from allowing greater flexibility in manufacturing (Hasan 2019). The need for physical prototypes is markedly reduced, accelerating the time taken from the ideation stage to actual product launch in the market. The application of concurrent engineering principles will enable faster and accurate execution of new designs into production, which inevitably gives the furniture manufacturers a larger product diversity. In essence, quality systems and technology application have transformed the global furniture manufacturing industry, by minimizing trails and errors to improve productivity, and at the same time, blending the art and science of furniture design and manufacturing more effectively.

Summary

- Furniture engineering is aimed toward making the furniture shape has the important strength and stiffness to carry out its function.
- It calls for the utility of material science and structural mechanics knowledge, not only for the strength design however also the overall performance reliability of the shape.
- Although furniture engineering has advanced significantly, through the application of FEM techniques, physical prototypes are nevertheless required for final client approval.
- The performance testing of furniture is obligatory in many jurisdictions around the world, and standards have to be complied with when carrying out those assessments scientifically.

References

Anon (1989) Encyclopedia of wood. Revised edn. Sterling Publication Inc., New York, USA

Anon (2014) Safety, performance and environmental testing of commercial and residential furniture. UL LLC Publication, Northbrook, Illinois, USA

Anon (2015) Furniture testing 101—a definitive guide to North America furniture testing. Intertek Publication, London, UK

Bao Z, Eckelman C, Gibson H (1996) Fatigue strength and allowable design stresses for some wood composites used in furniture. Holz Roh Werkst 54(6):377–382

Buckingham R (2009) Engineer's guide for evaluating furniture quality and improving test methods. Intertek Publication, London, UK

Case J, Chilver L, Ross CTF (1983) Strength of materials and structures, 3rd edn. Arnold Press, London, UK

Chandrupatla TR, Belegundu AD (1991) Introduction to finite elements in engineering, 2nd edn. Prentice-Hall Inc., Upper Saddle River, NJ, USA

Durka F, Morgan W, Williams DT (1996) Structural mechanics. Addison-Wesley Longman Ltd., Essex, UK

Eckelman CA (1978) Strength design of furniture. Timber Tech. Inc., West Lafayette, Indiana, USA

Eckelman CA (2003) Textbook of product engineering and strength design of furniture. Purdue University Press, West Lafayette, Indiana, USA

Erdil YZ, Haviarova E, Eckelman CA (2004) Product engineering and performance testing in relation to strength design of furniture. Wood Fiber Sci 36(3):411–416

Hasan EFE (2019) 'Basic design is the mother of design' and furniture design. In: Proceedings of the 29th international conference on research for furniture industry, 19–20 September 2019, Ankara, Turkey, pp 279–302

Hearn EJ (1997) Mechanics of materials, vol II, 3rd edn. Butterworth-Heinemann, Oxford, UK

Ho CL, Eckelman CA (1994) The use of the performance tests in evaluating joint and fastener strength in case furniture. Forest Prod J 44(9):47–52

Hu W, Liu N, Guan H (2019) Optimal design of a furniture frame by reducing the volume of wood. Drewno 62(204):85–97

John VB (1992) Testing of materials. Macmillan Press Ltd., London, UK

Prekrat S, Pervan S, Smardzewski J (2011) Optimization of furniture testing. Annals of Warsaw University of Life Sciences—SGGW—Forestry and Wood Technology, no 73, pp 60–65

Ratnasingam J, Ioras F (2010) Static and fatigue strength of oil palm wood used in furniture. J Appl Sci 10(11):986–990

Ratnasingam J, Ioras F (2011a) Bending and fatigue strength of mortise and tenon furniture joints made from oil palm lumber. Eur J Wood Wood Prod 69(4): 677–679

Ratnasingam J, Ioras F (2011b) Fatigue strength and design stress of oil palm wood for furniture application. Eur J Wood Wood Prod 69(3): 507–509

Ratnasingam J, Ioras F (2013) Effect of adhesive type and glue-line thickness on the fatigue strength of mortise and tenon furniture joints. Eur J Wood Wood Prod 71(6):819–821

Ratnasingam J, Ioras F (2015) The fatigue characteristics of two-pin moment-resisting dowel furniture joints with different assembly time and glue-line thickness. Eur J Wood Wood Prod 73(2):279–281

Ratnasingam J, Mutthiah N (2017) Fatigue life of oil palm wood (OPW) for furniture applications. Eur J Wood Wood Prod 75(3):473–476

Ratnasingam J, Perkins M, Reid H (1997) Fatigue: Its relevance to furniture. Holz Roh Werkst 55(5):297–300

Ratnasingam J (2019) Furniture testing and its implications on strength design of furniture. Tech. Note, no 16, IFRG, Singapore

Simanová L, Gejdoš P (2015) The use of statistical quality control tools to quality improving in the furniture business. Procedia Econ Financ 34:276–283

Šimek M, Sebera V (2015) Digital image correlation in furniture testing. In: Proceedings of the in Wood 2015 conference—inovations in wood materials and processes, Brno, Czech Republic, 19–22 May 2015, pp 86–87

Smardzewski J (1998) Numerical analysis of furniture construction. Wood Sci Technol 32:273–286

Smardzewski J (2015) Furniture design. Springer International Publishing, Cham, Switzerland

Thompson RJH, Ansell MP, Bonfield PW et al (2002) Fatigue in wood-based panels. Part 1: The strength variability and fatigue performance of OSB, chipboard and MDF. Wood Sci Technol 36(3): 255–269

Vickery SK, Dröge C, Markland RE (1997) Dimensions of manufacturing strength in the furniture industry. J Oper Manag 15(4):317–330

Zhang J, Quin F, Tackett B (2001) Bending fatigue life of two-pin dowel joints constructed of wood and wood composites. Forest Prod J 51(10):73–78

Zhang J, Li G, Sellers T Jr (2003) Bending fatigue life of two-pin dowel joints in furniture grade pine plywood. Forest Prod J 53(9):33–39

Chapter 13
Automation Technology in Furniture Manufacturing

The increased application of automation and technology in the furniture manufacturing industry is strongly advocated as solution to minimize its dependency on manual workforce. Being a labor-intensive industry, furniture manufacturing lag in terms of investments in technology compared to other manufacturing industries. Nevertheless, the adoption of low-cost automation technologies, especially the use of pneumatic and hydraulic tools, is increasing within the industry, and will inevitably assist in improving productivity and throughput rate. Further, digital technologies in the realms of marketing activities are also increasing worldwide. As the adoption of Industry 4.0 technologies picks up pace globally, the furniture manufacturing industry must also prepare to embrace this technological revolution, which offers many benefits, in almost all aspects of furniture manufacturing.

13.1 Introduction

The furniture manufacturing industry in many parts of the world, which started off as a cottage-based industry, grew rapidly to become a large volume manufacturer of furniture, by leveraging on the concept of economies of scale. The idea of mass-production, although extending cost competitiveness, limit the ability of the manufacturer to have product diversity, production value flexibility, and importantly the extent to which value-addition takes place. Although sub-contracting and outsourcing practices are widespread throughout the world, its main motivation remains to be gaining cost competitiveness (Ratnasingam 2020).

The furniture customers become more demanding customization of furniture manufacturing is increasingly common in many countries. Customers want greater choices and selections in terms of fittings, fixtures, materials, fabrics, etc., and yet the furniture must be priced affordable (Foster 2019). Furniture manufacturers are faced with the challenges of coping with such demands, and therefore, the existing manufacturing systems may no longer be appropriate. Further, with globally growing

J. Ratnasingam, *Furniture Manufacturing*, Design Science and Innovation,
https://doi.org/10.1007/978-981-16-9412-7_13

e-commerce trade and *digitalization*, the furniture manufacturing industry needs to be adaptive to cope with such requirements.

It is indeed timely for furniture manufacturers to seriously look at investing in automation and technologies that will help transform their industry, instead of depending on skilled workmen and carpenters, whose availability is reducing throughout the world. The application of *automation* and *Industry 4.0* offers promising solution to furniture manufacturers, not only to manufacturer a variety of products, but also in small volumes, at an affordable cost. The concept of *smart factories* allows optimized production flow, while permitting small production batches, with versatile designs (Machado and Davim 2020). In such smart factories, the designs are created digitally, based on the customer's inputs, transferred to the respective machines and workstations, to be manufactured quickly. In the end, the parts, components, etc. are collated and packed for distribution. This scenario may appear far-fetched, but it is on the horizon and will soon be part of the furniture manufacturing industry.

13.2 Automation

Automation is the process in which automated instruments and devices, monitor and control functions that were previously done by humans. The primary aim is to make the best use of all resources, apart from increasing labor efficiency and product quality (Frohm et al. 2008).

Factors determining the degree of automation in any manufacturing outfit are the costs involved and its feasibility in the existing production environment, including skilled workers, computer literacy, manufacturing system, scale of production, etc. Contrary to common belief, automation does not replace the workforce completely, but the workforce acquires new qualitative nuances through retraining, which turns their jobs into more meaningful. In most circumstances, the humans will provide technical support to the automated machinery, apart from carrying out analytics and administrative activities.

The many repetitive and low-skilled dependent operations in furniture manufacturing are the early targets for automation. With a cost-sensitive manufacturing environment, the automation solutions offered must be simple, reliable, and efficient to perform specific repetitive functions, while achieving the necessary monitoring and control of the operation through the computer-machine language interphases (Landscheidt and Kans 2016).

Pneumatic and *hydraulic* systems are integral parts of automation in the furniture manufacturing industry, which involves multiple repetitive, mechanical motions. These systems minimize, if not eliminate the errors during the monotonous operations (Daines and Daines 2018; Groover 2018). The ***pneumatic system*** uses compressed air to regulate actuating mechanisms in most woodworking machines. Apart from that, clamping devices, spray application, nail guns, etc. also use compressed air for its operation and control. Since gases are highly compressible, light, and with no specific volume, pneumatic systems are simply low-cost automation (LCA) solutions that have found extensive application in furniture manufacturing technology.

On the other hand, the ***hydraulic system*** transmits a more consistent and higher power to the actuating mechanism, compared to the pneumatic system. It offers the advantage of providing smooth, flexible, uniform motion free from vibration, and unaffected by any load changes. Hence, the hydraulic systems are common for heavy-duty applications, such as press, shipping jacks, and conveyor systems, which require consistent and accurate power. Table 13.1 shows the major differences between the hydraulic and pneumatic systems.

Three types of automation can be distinguished in the manufacturing sector: (1) fixed automation, (2) programable automation, and (3) flexible automation.

Fixed automation refers to an automated manufacturing facility in which the sequence of manufacturing processes is determined by the equipment setup. In effect, the programmed instructions are stored in the machines as cams, gears, and other hardware, which does not allow easy changeovers from one product to another. Such automation requires large initial investment and high manufacturing costs. In the case of furniture industry, such automation is suitable for large volume production facilities (Johansson et al. 2016).

Table 13.1 Comparison of hydraulic and pneumatic systems

Hydraulic system	Pneumatic system
Pressurized oil as medium to transmit power	Pressurized air as the medium to exchange power
High pressures of up to 70 N/mm^2	Low pressures between 0.5 and 1.0 N/mm^2
Design for closed system	Designed for open system
System slows, if leakage occurs	System not influenced by leakage
Valve operation is difficult	Valve operation is easy
Heavier in weight	Lighter in weight
Pumps required to pressurize liquid	Compressors are utilized to supply compressed air
Automatically lubricates	Special lubrication required
Limited speed	High speed possible
No noise produced	Noisy
Resistant to fluctuating load	Affected by fluctuating load
More expensive	Less expensive
Suitable for feed movement	Not appropriate for feed movement
Cavitation a serious problem	No cavitation problem

Source Groover (2018)

Programmable automation refers to a type of automation that is used to manufacture products in batches. Every new batch requires reprogramming and conversion of the manufacturing system to accommodate the new product, which is time consuming, and often leads to production downtime. However, the production rate in such automation system is much lower, as it allows product changeovers. A good example of programmable automation is numerical control (NC) machines, where the program for each product type is coded into the computer, which then operates the machine tools. In the manufacturing of 32 mm series flat-panel furniture, such as case-tops, such automated systems are common (Ratnasingam 2020).

Flexible automation allows a variety of product designs to be manufactured, with quick changeovers taking place automatically through the control of computers. The programs for the various product types are coded into the computer offline, which significantly reduces downtime, but the production rate is compromised. Such automation system can cope with very variable product types, and hence, has potential for application in solid-wood furniture manufacturing (Groover 2018).

An *automated production line* is made up of a series of workstations connected by a transfer system to move components and parts from one station to the other. Such a system is commonly found in factories manufacturing flat wood-based panels and metal furniture (Landscheidt et al. 2017). This is an example of the fixed automation system, where every workstation performs specific tasks, adding incremental value to the components and parts, as they move along the production line. These operations are controlled and coordinated by programmable logic controls, in which computers interface with machine workstations, to time and sequence every operation precisely to ensure and efficient production flow. Tool changes, machine setup, etc. are automatically controlled, which in turn significantly reduces downtime, while production rates are kept high (Foster 2019). This is a stark contrast to the previous practice of scheduling and sequencing the various operations manually through the factory shopfloor, which is dependent on skilled and knowledgeable production planning and control (PPC) personnel to execute an efficient operation.

Previously, numerical control was the most widely employed automation technology. Numerical control (NC) is a type of programmable automation in which a system is managed by means of numbers (and other symbols) encoded on punched paper tape or another type of storage medium. Numerical control was commonly used in the machine-tool industry to control the position of the workpiece being machined relative to the cutting tool. The NC software is a collection of machining commands for a single part, or group of components (Mittal et al. 2018). The coded numbers in the software specify x-y-z coordinates in a Cartesian axis system, defining the various positions of the cutting device relative to the position of the workpiece. By sequencing the movements of the cutting tool, which is monitored by a position feedback control system, the machining operations are accomplished efficiently.

When a computer is attached to a NC machine as a controller, the operational program is actuated from the computer, rather than the punched paper tape. Under such circumstances, the machine is known as a *computer numerical control* (CNC) machine. The CNC workstation performs multiple machining functions and has found widespread application in the furniture manufacturing sector (Ratnasingam 2020).

Assembly operations within the furniture enterprise have generally been handled manually and are a labor-intensive operation. Automated assembly lines are suitable in high volume production environments, where the products are small with simple designs. Unfortunately, in many solid-wood furniture factories, these criteria are not met, necessitating manual assembly sections (Ratnasingam et al. 2019).

Automated assembly machines function in a similar manner as the machining transfer lines, where several stations feed a specific component or part for the final assembly, and the work-head at the assembly station performs the attachment of all components/parts into the final assembly. In fact, a component or part is added on to the product as it moves along the line until the final assembly is completed. Such assembly lines are considered fixed automation because of the higher volume production it can handle. On the other hand, if programable automation is used, as in the case of dowel inserting machines, multiple product design is an option (Frank et al. 2019).

Perhaps, *robotics* is the fastest growing form of automation across the global manufacturing industries (Landscheidt et al. 2017). The application of robotics in the manufacturing sector is common for: (1) material handling, (2) processing operations, and (3) assembly and inspection.

Material-handling applications include material transfer and machine loading and unloading. In a production line, the robots carry out loading and unloading of components and parts from the various conveyor systems, using the gripper attached to its arm, which could handle different part geometry. In complex operations, the robots may be involved in stacking up packaged furniture on a pallet, while keeping count of the inventory at the same time (Büchi et al. 2020).

When the robot manipulates a device or a tool, to perform a task on a workpiece, the robot is said to be involved in *processing operations*. An example of this is the use of robotic arms to carry out automatic spraying operations in the factory. Other processing operations that could be handled by robots include sanding, polishing, and routing.

The third application area of industrial robots is in *assembly and inspection*. Traditionally, these operations involve quite a lot of manpower, and hence, the use of robots for such operations is growing worldwide. Since robots are programable, the assembly of multiple products is easily handled. With appropriate sensors, robots can be programmed to carry out inspection to ensure consistent component quality, while at the same time, picking up defective ones.

Robotics provides an alternative to manpower in practically all manufacturing sectors, especially when: (1) the job is repetitive, monotonous, and involving the same mechanical motions in each cycle; (2) the activity is hazardous, or unpleasant for the humans (e.g., spray painting, sanding, and material handling); (3) the job involves using a tool that is heavy and difficult to handle; and (4) the production needs to run throughout the day (Groover 2018).

The significantly higher production rates and improved productivity levels are the two main reasons that favor the application of automation in the manufacturing sector. Other advantages offered include more consistent quality, reduced downtime on the factory shopfloor, improved safety, and reduced wastage. (Chui et al. 2017). Indirectly, automation reduces the workweek for the labor force.

Employee displacement is the most common drawback of automation. The large initial capital expenditure, higher maintenance required, and reduced flexibility are the other challenges faced when considering automation (Chui et al. 2017). It appears that these challenges remain a major concern for the SMEs predominated furniture industry in many parts of the world. In fact, furniture manufacturing industry lag the overall manufacturing sector in terms of application of automation and robotics. For instance, the furniture industry in the Southeast Asian region deploys 8 robots per 10,000 employees, as opposed to the 90 robots per 10,000 employees in the overall manufacturing sector (Ratnasingam et al. 2019).

13.3 Digital Technology

The globalization of the furniture industry has forced many manufacturers to pay more attention to digital technologies, not only to reduce the lead time from product design to production, but also to expand the market share through e-commerce and digital marketing. As a matter of fact, the application of digitalization and related technologies has increased throughout all sectors of the global economy, since the Covid-19 global pandemic struck in late 2019. The cancellation of major furniture fairs and other marketing shows throughout the world, had forced traditional furniture marketeers to adopt digital marketing tools that were becoming sophisticated, and effective. Digital technologies involved in operations cover a wide array of technologies, some of which has been described previously. These include computer numerical control (CNC) machining centers, computer-aided design (CAD), computer-aided-manufacturing (CAM), and computer-integrated-manufacturing (CIM). In fact, digital technologies are also finding applications in supply chain management, inventory control, and purchasing and procurement functions in the furniture manufacturing industry (Robb et al. 2008).

E-commerce and digital marketing technologies are increasingly important at the moment due to the Covid-19 pandemic, but the trend is expected to continue, if not accelerate in the post-pandemic era. Among the common Internet-based marketing tools are websites, emails, search engine optimization (SEO), *Google Adwords*, newsletter, article marketing, banner marketing, *YouTube* channels, and social media marketing. Social-media marketing has also rapidly expanded with the emergence of many different platforms, including *WhatsApp*, *LinkedIn*, *Facebook*, *Instagram*, *Twitter*, *Skype*, *Myspace*, *Reddit*, *Google Plus*, etc., which aims to seek, connect, engage, and build a relationship not only with new customers, but also existing customers (Fig. 13.1). Digital marketing is the fastest growing service within the global furniture industry, as the contemporary furniture marketing channels such as exhibitions, retail malls, specialty stores, and galleries are no longer viable in the new norm, where online digital technologies are the predominant tool for marketing

Fig. 13.1 Common digital platforms

activities (Ratnasingam et al. 2021a, b). In fact, hybrid marketing channels is here is stay for the long term.

13.4 Industry 4.0

Industry 4.0 refers to the Fourth Industrial Revolution and represents a new level of management and control of industrial value-chains (Sander et al. 2016). It is characterized by a high level of data exchanges from automation and production technology systems, through cyber-physical network, Internet of Things (IoT), cloud computing, and cognitive computing, leading to a *smart factory* (Zheng et al. 2018). The features of Industry 4.0 include (i) *even more automation than in the third industrial revolution,* (ii) *the bridging of the physical and digital world through cyber-physical systems, enabled by Industrial IoT,* (iii) *a shift from a central industrial control system to one where smart products define the production steps,* (iv) *closed-loop data models and control systems* and (v) *personalization/customization of products* (Schwab 2018).

In a nutshell, Industry 4.0 is the facts-extensive transformation of manufacturing (and related industries) in a connected environment of big data, people, processes, services, and systems in an IoT-enabled ecosystem. In such ecosystems, data and information are generated, leveraged, and utilized to carry out tasks in the smart factory, which is innovative and collaborative in nature (Zhang et al. 2017) (Fig. 13.2).

The key enabling technologies (KETs) of Industry 4.0 is described in Table 13.2, and its important application areas identified (Büchi et al. 2020; Misra et al. 2020). In essence, Industry 4.0 provides a wide range of technologies that could transform furniture design and production processes. Nevertheless, previous studies have shown that the application of one, several, or all of these technologies will be determined by the costs involved, skills availability, required capacity, and the prevailing level of information and computers technology (ICT) application in the manufacturing environment (Dalenogare et al. 2018; Mayropoulos and Nilsen 2020). This is of

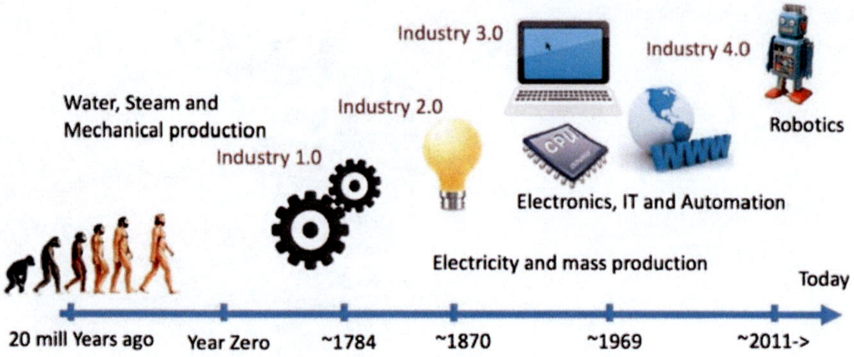

Fig. 13.2 Timeline for technological evolution

particular importance as the furniture manufacturing industry is predominated by SMEs throughout the world, and its level of investments in computing and automated technologies remains relatively low. Research has also shown that investments in technologies is generally higher among large-sized enterprises compared to SMEs in furniture manufacturing (Ratnasingam et al. 2019).

The adoption of Industry 4.0 in the furniture industry, would transform the industry from the current *flexible manufacturing facilities* into *smart factories* (Murmura and Bravi 2018; Landscheidt and Kans 2019).

Flexible manufacturing systems (FMS), which is widely adopted in the furniture industry, are based on the concept of flexible automation, where several workstations, linked together by conveyors or material-handling system are controlled by a computer system. The distinguishing difference between the FMS and the automated production facility is the ability of the former to manage several different product types simultaneously. Further, FMS allows changes in product mix and volume production and therefore is suitable for manufacturers with low to medium production volumes (Bambura et al. 2020). The characteristics of the FMS are that: (1) most of the workstations are CNC, or with a high degree of programmable automation, (2) a good conveyor or material-handling system linking every workstation is in place, (3) a significant computer network is installed to coordinate, manage, and control the machining and material-handling systems, and (4) skilled workforce is available to maintain and support all mechanical and computer systems (Frank et al. 2019).

Since the late 1990s, the global furniture manufacturers have adopted computers to facilitate product design and manufacturing activities. The concepts of computer-aided design (CAD) and computer-aided manufacturing (CAM) brought about major shift in design and manufacturing efficiencies. In the last decade, computer application has been extended to other functions within the furniture factories. This manufacturing concept is known as computer-integrated manufacturing (CIM) (Groover 2018).

Table 13.2 Key enabling technologies of Industry 4.0

Enabling technologies		Definition	Opportunities	Application
1	Advanced manufacturing Solutions	This incorporates interconnected and modular solutions for manufacturing facilities. Examples include automatic materials handling, robotics, and low-cost automation solutions	– Reductions in setup costs, and downtimes – Flexible production through workers' participation – Higher factory throughput	Machining centers; Conveyorized factory-floor
2	Augmented reality	This uses mobile devices/gadgets to enhance human sensory perceptions, such as sound, smell, and touch, or the virtual environment	– Higher speed in prototyping with increased virtual reality – Improved quality and less waste	Product development; improved customer experience
3	Internet of Things	This involves a set of devices and sensors that allows communication between people, machines, and products	– Allows higher product quality based on customer needs – Easier to trace and track product manufacturing, origin, shipment, and use – Greater interconnection along the supply and distribution network allowing better customer service	Global supply chain; Market research; Production system
4	Big data analytics	This involves analyzing large volume of data and information derived from products, processes, machines, and people	– Expedited communication with customers about the products – Flexible production capacity – Optimized supply chain	Global supply chain; Consumer Preferences; Market Research

(continued)

Table 13.2 (continued)

Enabling technologies		Definition	Opportunities	Application
5	Cloud computing	It archives and processes large quantities of data and information at high speed and efficiently	– The opportunities and challenges are similar to Big Data Analytics and Internet of Things	Global supply chain; Marketing; E-Commerce
6	Cyber security	This consists of security measures/software installed to protect the flow of data and information through the Internet	– The safety and protection of data and information between interconnected technologies	Copy-right Computing systems protection
7	Additive manufacturing	This additive production technique allows for complex products to be made by incremental layering of material, as in the case of 3D printing	– Higher speed in prototyping – High product quality with less waste – Reduces cost	Product development; Prototype making manufacturing process improvements
8	Simulation	This involves reproducing the physical world in virtual platforms, which allows the user to optimize all aspects of the product and its production	– Higher speed in prototyping and product development – Significant savings in downtimes	Assessment of consumer preferences and behavior; Product development
9	Horizontal and vertical integration	The integration offered by Industry 4.0 is characterized by way of two dimensions: internal/horizontal and external/vertical. Examples include the material requirement planning (MRP) and enterprise requirement planning (ERP) systems	– Significant improvements in production planning and control, cost reduction and production quality – High productivity and capacity utilization	Production management and control Supply chain management;
10	Other enabling technologies	Technologies incorporating artificial intelligence (AI) and sensors improve production control, while reducing waste on the factory shopfloor	– Optimizes the production flow and reduces cost	Production management and control

Source Büchi et al. (2020)

The CAD/CAM system is predicated on the computer's ability to prepare, store, and display large amounts of data related to the products and its components. This may include the *bill of materials* (BOM) , graphical models of the products, components specifications, etc. Therefore, the CAD/CAM system automates the entire design to manufacturing phases.

The CAD allows the creation, analysis, and optimization of the product design, without even requiring the designer to work on the drafting board, making sketches, line drawings, etc. The computer system would display the product, allowing design changes to be made, apart from carrying out structural analysis by a supporting software. At the end, the CAD system will produce a detailed drawing of the product, after the adjustments, ready for manufacturing (Abu et al. 2019).

The use of computer systems in the planning, control, and administration of manufacturing operations is known as computer-aided manufacturing (CAM). This can be fulfilled through direct or indirect linkages between the computer and the manufacturing functions. In direct linkage configurations, the computer is used to monitor, control, and manage the manufacturing functions, and also providing feedback to the factory management on the process performance. Such a system will improve manufacturing productivity (Machado and Davim 2020).

Computers enhance production efficiency by supporting the various processes, without controlling and managing it. In such indirect linkages, the computer supports the production planning function, which becomes more efficient, compared to what is done by humans only. These may include functions such as step-by-step planning for the production sequence, numerical control programming, and scheduling the operations in the factory. Research has shown that manufacturing efficiency improves by at least 12% under such circumstances (Murmura and Bravi 2018).

The CIM concepts encompass all engineering and business operations of the furniture manufacturing factory. These will include business functions, such as inventory and purchasing management, cost accounting, human resource and payroll management, sales and involving, as well as factory administration. On the other hand, engineering functions ranging from product design and development right through manufacturing, and production planning and control are managed by the CIM system. In typical CIM systems, all operations from the point of customer order to shipping the finished products are handled (Säfsten et al. 2007). CIM is considered an intelligent factory, where all manufacturing-related activities are controlled by computers (Masood and Sonntag 2020). In fact, furniture factories that have adopted the CIM system, are steps away from shifting towards Industry 4.0 (Mittal et al. 2018).

In conclusion, automation, digitalization, and industry 4.0, apart from transforming the labor-intensive manufacturing environment in the furniture industry, offers a completely new approach to furniture manufacturing (Simonová 2015; Abu et al. 2019; Ratnasingam 2020). By embracing innovations, and transformed business processes, models, and strategies, the furniture manufacturing industry is poised to make huge steps in terms of increasing profits, reducing cost, and enhanced value-addition. This in turn will help boost customer loyalty, increase sales through value creation, and most importantly become a sustainable industry. In order to make this transformation a reality, it is necessary to approach furniture manufacturing from a production engineering perspective, rather than simply a craft-based industry.

Summary

- Automation, digitization, and Industry 4.0 technologies offer the furniture industry to increase productivity and move up further the value-chain.
- Nevertheless, the application of such technologies may incur higher investment cost, better human capital with the necessary skills set and knowledge, and a well-networked manufacturing facility that is data and information driven.

References

Abu F, Gholami H, Mat Saman MZ et al (2019). The implementation of lean manufacturing in the furniture industry: a review and analysis of the motives, barriers, challenges, and the application. J Clean Prod 234: 660–680

Bambura R, Sujovâ E, Ĉierna H (2020) Utilizing computer simulation to optimize furniture production system. BioResources 15(3):6752–6765

Büchi G, Cugno M, Castagnoli R (2020) Smart factory performance and industry 4.0. Technol Forecast Soc 150: 119790

Chui M, Manyika J, Miremadi M et al (2017) Human + Machine: a new era of automation in manufacturing. McKinsey & Co. Publication, Munich, Germany

Daines JR, Daines MJ (2018) Fluid power—hydraulics and pneumatics, 3rd edn. Goodheart-Willcox Inc. Publication, Tinley Park, Illinois, USA

Dalenogare LS, Benitez GB, Ayala NF et al (2018) The expected contribution of Industry 4.0 technologies for industrial performance. Int J Prod Econ 204:383–394

Frank AG, Delenogare LS, Ayala NF (2019) Industry 4.0 technologies: implementation patterns in manufacturing companies. Int J Prod Econ 210:15–26

Frohm J, Lindström, Stahre, J, Winroth M (2008) Levels of automation in manufacturing. Ergonomia—Int J Ergon Human Factors 30(3): 1–28

Foster R (2019) Automation, production systems and computer-integrated manufacturing. Larsen and Keller Education Group, New York, USA

Groover M (2018) Automation, Production systems and computer-integrated manufacturing, 5th edn. Pearson Education Group, London, UK

Johansson J, Blomquist L, Nilson H et al (2016) Influencing factors to enable automation of the wood furniture production. In: Proceedings of the 12th meeting of Northern European network for wood science & engineering, Riga Latvia, pp 208-213

Landscheidt S, Kans M (2016) Automation practices in wood products industries—lessons learnt, current practices and future perspectives. In: Proceedings of the 7th Swedish production symposium, 25–27 October 2016, Lund, Sweden, pp 39–45

Landscheidt S, Kans M, Winroth M (2017) Opportunities for robotic automation in wood product industries—the supplier and system integrators' perspective. Procedia Manufactur 11:233–240

Landscheidt S, Kans M (2019) Evaluating factory of the future principles for the wood products industry: Three case studies. Procedia Manufactur 38:1394–1401

Machado C, Davim JP (2020) Industry 4.0: challenges, trends and solutions in management and engineering. CRC Press, Boca Raton, Florida, USA

Masood T, Sonntag P (2020) Industry 4.0: adoption, challenges, and benefits for SMEs. Comput Ind 121: 103261

Mavropoulos A, Nilsen AW (2020) Industry 4.0 and circular economy—towards a wasteless future or wasteful planet. John Wiley & Sons Inc., Hoboken, New Jersey, USA

Mittal S, Khan MA, Romero D et al (2018) A critical review of smart manufacturing and Industry 4.0 maturity models: implications for SMEs. J Manuf Syst 49:194–214

Misra S, Roy C, Mukherjee A (2020) Introduction to industrial Internet of Things and industry 4.0. CRC Press, Boca Raton, Florida, USA

Murmura F, Bravi L (2018) Additive manufacturing in the wood furniture sector: sustainability of the technology, benefits, and limitations of adoption. J Manuf Technol Manag 29(2):350–371

Ratnasingam J (2020). Status of automation and industry 4.0 adoption in the ASEAN wood products industry. Tech. Note, no 4A, IFRG, Singapore

Ratnasingam J, Ab Latib H, Lee YY et al (2019) Extent of automation and readiness for Industry 4.0 among Malaysian furniture manufacturers. Bioresources 14(3): 7095–7110

Ratnasingam J, Ioras F, Lim CL et al (2021a) Digital technology application among malaysian value-added wood products manufacturers. Bioresources 16(2): 2876–2890

Ratnasingam J, Natkuncaran J, Hazirah AL et al (2021b) Digital marketing during the COVID-19 pandemic: a case study of its adoption by furniture manufacturers in Malaysia. Bioresources 16(2): 3304–3317

Robb DJ, Xie B, Arthanari T (2008) Supply chain and operations practice and performance in Chinese furniture manufacturing. Int J Prod Econ 112(2):683–699

Sanders A, Elangeswaran C, Wulfsberg JP (2016) Industry 4.0 implies lean manufacturing: research activities in Industry 4.0 function as enablers for lean manufacturing. J Ind Engineering Manag 9(3): 811–833

Simanová Ľ (2015) Specific proposal of the application and implementation Six Sigma in selected processes of the furniture manufacturing. Procedia Econ Financ 34:268–275

Säfsten K, Winroth M, Stahre J (2007) The content and process of automation strategies. Int J Prod Econ 110(1/2):25–38

Schwab K (2018) Shaping the fourth industrial revolution. World Economic Forum Publication, Geneva, Switzerland

Zhang RY, Xu X, Klotz E et al (2017) Intelligent manufacturing in the context of Industry 4.0: a review. Engineering 3(5): 616–630

Zheng P, Wang H, Sang Z et al (2018) Smart manufacturing systems for Industry 4.0: conceptual framework, scenarios, and future perspectives. Front Mech Eng 13:137–150

Chapter 14
Cost Optimization in Furniture Manufacturing

The concept of optimization is being widely adopted throughout the furniture manufacturing industry in the world, both for better cost and quality control. In this context, the engineering approach to the many value-adding processes in furniture manufacturing is a necessity, as without accurate process engineering data on the production rate of each, and every process on the furniture factory shopfloor, achieving optimum production may never be realized. This chapter is the first attempt to bring both engineering and mathematical approaches to the determination of production rate in furniture manufacturing.

14.1 Introduction

In many instances, the furniture industry is considered a craft-based industry, and therefore lacks the application of engineering and mathematical principles, in its operations management. In fact, the furniture manufacturing processes often precludes precise calculations to optimize the process and is usually subjected to guestimates, which leads to inefficiency and wastages. Inevitably, the furniture manufacturing process will be more expensive to operate, and equipment will wear out prematurely. Consequently, the product quality suffers.

J. Ratnasingam, *Furniture Manufacturing*, Design Science and Innovation,
https://doi.org/10.1007/978-981-16-9412-7_14

In order to minimize this negative impact, if not eliminate it all together, calculations to measure raw materials consumption, machining outputs, and production cost must be carried out accurately on the basis of established scientific and engineering principles. This will allow the furniture manufacturing process to be continuously monitored, controlled, and optimized for the best outcome. Generally, optimization of the furniture industry involves the following steps: (1) optimize the process, (2) reducing the material costs, (3) improve workforce productivity, and (4) control the inventory carrying costs (Ratnasingam 2015). An optimized production process will help avoid over-production and reduce excess storage costs, and it involves an accurately controlled production cost through a mathematical approach.

Traditionally the application of engineering concepts to furniture manufacturing is limited, as the trade had always assumed a craft approach. In this context, there is much emphasis on experience gained through trial and error, rather than objective engineering principles. Such an approach may have been reliable in the past, but with increasing application of machine technology in furniture manufacturing, the approach to furniture manufacturing must take the engineering perspective in order to ensure a productive operation (Ratnasingam 2015).

14.2 Production Engineering Approach to Raw Materials Use

One of the major challenges faced in the optimization of furniture manufacturing processes is the inability to accurately ascertain raw materials use and its yield/recovery during processing. The wood materials, usually in the form of sawn timber, are often traded in the imperial system of measure, while in the furniture manufacturing operations, measurements are made in the metric system. Further, when dealing with imported wood species, the board foot measurement units are used. The purpose of the following formulas is to provide guidelines on how these conversions and calculations are carried out in industrial practice.

Wood Material	Wood-Based Panels
$1.416\ m^3 = 1$ Ton $= 50\ ft^3$ 1 ft = 12 inches 1 inch = 25.4 mm 1000 board foot = $2.36\ m^3$	12 mm x 8 feet x 4 feet = $0.035\ m^3$ 18 mm x 8 feet x 4 feet = $0.053\ m^3$

Adhesive

Volume of adhesive (m^3) x Specific Gravity (SG) = Weight of adhesive required (kg)

Bond-line thickness (mm) x Total surface area of bond-line (mm^2) = Volume of adhesive (m^3)

$1\ m^3 = 1000$ liter

Bond-line (mm)	*Theoretical area (m^2) covered by 1 L of adhesive*
0.05	*20*
0.15	*7*
0.20	*5*
0.25	*4*
0.40	*2.5*

Note: SG of the adhesive is usually provided in the technical data sheet of the adhesive from the supplier

Finish Materials

Wet film thickness (μm) = Dry film thickness (μm) ÷ Volume solids (%) x 100

Dry film thickness (μm) = Wet film thickness (μm) x Volume solids (%) x 100

Volume solids = % x 100

Spreading rate (m^2/l) = Volume solids (%) x 10 ÷ Required dry film thickness (μm)

$$\text{Volume of coating used (l)} = \frac{\{10 \times \text{Area to be coated } (m^2) \times \text{Dry film thickness } (\mu m)\}}{\{\text{Volume solids } (\%) \times (100 - \%\ \text{wastage}\}}$$

Effect of solvent/thinner on volume solid of wet film thickness

$$\text{Corrected volume solid } (\%) = \frac{\{\text{Volume solid } (\%) \times 100\}}{\{100 + \%\ \text{thinner added}\}}$$

Corrugated Carton Box

Width (cm) x Length (cm) x Thickness (cm) = Volume (cm^3).

$1\ m^3 = 1{,}000{,}000\ cm^3$

Note: external dimension of the carton box is taken into consideration, while the internal dimension is calculated after deducting the wall thickness (i.e., usually 4–7 mm).

Understanding the mathematics involved in converting and calculating the materials use in the furniture manufacturing process would enable standard consumption and cost values to be established so that the necessary control measures can be implemented should there be variance on the shopfloor.

14.3 Production Engineering Concepts in Furniture Manufacturing

In machining operations, several machining parameters must be established in order to determine the production capacity of the machine, its production cost and the power consumption. Further, such a mathematical approach brings an engineering perspective to the machining processes involved in furniture manufacturing and will assist towards optimizing the machining process, to achieve process economics as well as quality standard (Ratnasingam 1999).

In furniture manufacturing, the quality and direction of cutting the wood materials has a strong influence on the process yield or recovery. Being an anisotropic material, wood has three unique planes, i.e., longitudinal, radial, and tangential. Hence, in wood machining operations, in almost all circumstances, the direction of the cut would involve one or two of these surfaces resulting in non-optimal surface (Fig. 14.1). For the first approach, the deliberations below will assume the most mechanical machining operations in furniture manufacturing are peripheral milling operations.

In peripheral milling operations, the cutter makes intermittent engagement with the workpiece, in a series of *cycloid cutting circles* that overlaps each other, depending on the feed speed and the cutting speed. As the cutter bites into the workpiece, a quantity of material is removed by the cutting wedge leaving a series of marks on the surface, known as *cutter marks* (Fig. 14.2).

The quality of the machined surface is influenced, not only by the cutting parameters, but also the stock removal rate, which inevitably has an implication of the process economics.

Fig. 14.1 Directions of cut in wood machining

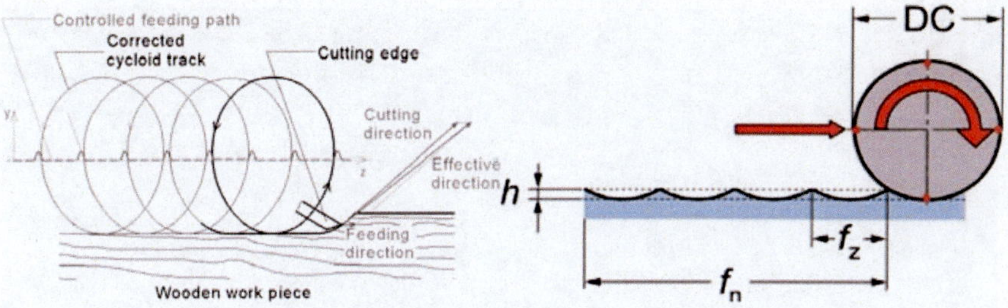

Fig. 14.2 Engagement of cutting circles and cutter marks on a work piece

Peripheral Cutting Speed	Feed Speed
$V = \frac{\pi \times D \times R}{1000 \times 60}$ V = cutting speed (m/s), π = 3.14, D = diameter of cutter (mm), R = revolution per minute of cutter (RPM)	$F = \frac{N \times R \times P}{1000}$ F = feed speed (m/min), N = number of cutting edges in cutter block, R = revolution per minute of cutter (RPM), P = pitch distance between cutter marks (mm)
Pitch/Distance between Cutter Marks	**Height of Cutter Mark**
$P = \frac{1000 \times F}{N \times R}$ P = distance between cutter-marks (mm), F = feed speed (m/min), N = number of cutting edges in cutter block, R = revolutions per minute of cutter (RPM)	$H = \frac{P^2}{4 \times D}$ H = height of cutter mark (mm), P = distance between cutter marks (mm), D = diameter of cutter block

Average Thickness of Chip $t = P\sqrt{\frac{d}{D}}$ t = average chip thickness (mm), P = distance between cutter marks (mm), d = depth of cut (mm), D = diameter of cutter block (mm)	**Out of Balance Force in Cutters** $F_c = \frac{M \times R\,(2 \times \pi \times N)^2 \times 10}{g \times 60^2}$ F_c = centrifugal force (N), M = off-balance mass of cutter (kg), R = radius of gyration of center of gravity of cutter in m, N = RPM, π = 3.14, g = gravitational force which is 10 kg.m/s^2
Material Removal Rate (MRR) $S = d \times w \times F$ S= Stock removal rate (mm^3/min), d = depth of cut (mm), w = width of cut (mm), F = feed speed (mm/min)	**Work Done during Machining** Work done = Kinetic energy expanded $W = E_k$ Power consumed (Horse-power, Hp)

If the mass of the stock removed is to be determined, the equation is transformed as: $M = d \times w \times F \times \rho$ M = mass of stock removed (mg/min), d = depth of cut (mm), w = width of cut (mm), F = feed speed (mm/min), ρ = density of material (mg/mm^3) **Chip Load per Knife** Chip Load, $C_p = F \div (R \times N)$ C_p = Chip load (mm/tooth), F = feed speed (mm/min), R = RPM, N = number of teeth	$H_p = a \times S^n$ Hp = kilowatt (kW) of power required, a = horse-power at the cutter when the rate of stock removal is 1mm^3/min, S = stock removal rate (mm^3/min), n = exponent of stock removed 1 Hp = 745 Nm/s = 745 Watts = 0.745 kW = 745 J/s Therefore, transforming the equation accordingly to determine the cutting power required. $P_c = \frac{d \times w \times V \times k_c}{60 \times 10^6}$ P_c = cutting power (kW), d = depth of cut (mm), w = width of cut (mm), F = feed speed (mm/min), kc = specific cutting force (N/mm^2) Average cutting force for wood is between 60 and 102 N/mm^2

In essence, by defining the process parameters and stock removal rate, the process economics can be mathematically determined, providing a benchmark value for comparison purpose.

14.4 Engineering Approach to Manufacturing Cost

In the furniture manufacturing industry, the manufacturing cost is the sum total of the machining cost, tooling cost and unit labor cost. Unlike conventional product cost, manufacturing cost focuses on the cost of the various machining processes involved in making the piece of furniture. In practice, the machining cost is a function of the production rate of the machines. The following formulas allow the calculations of the related costs to be ascertained.

Tooling Cost

Cost of new tool = x

Cost per grinding = y

No. of grinding possible per tool = z

Edge life per grind = s (m)

Cost overall = $x + y + z = \gamma$

Life time overall (m) per tool = (10 + 1) x $s = z$

Cost per meter production for lifetime of tool = $\gamma \div z$

Labor Cost

$$\text{Labor Rate (DLR) per Day} = \frac{(\text{daily rate x days worked a month}) + \text{fringe benefits} + \text{incentives}}{\text{No. of days worked a month}}$$

$$\text{Unit Labor Hour (ULH)} = \frac{\text{No. of items produced per month}}{\text{Total labor hours per month}}$$

Unit Labor Cost = ULH x (DLR ÷ Hours per Day)

Production Capacity

$$P = F \times \{(60 \times T) - (C \times W)\} \times K$$

P = machine production capacity (meters-run), F = feed speed (m/min), T = length of working shift (hour), C = idle time for machine due to tool change (min), W = number of tool change, K = continuity of feed due to workers' downtime (%)

Machine Cost

Cost of machine ($\$^1$) – Salvage Value ($\$^2$) = Depreciation Value ($\$^3$)

Cost of Machine per Day ($\$^4$) = $\$^3$ ÷ (Machine life (years) x 12 x 8)

Note: It is assumed that a working shift is 8 hours

Machine Cost per meter Output (\$) = $\$^4$ ÷ P(meters-run)

Against this background, it is apparent that machining cost is indirectly influenced by the throughput rate and the capacity utilization. Therefore, it is no surprise that in many mass-production environment, maximum capacity utilization is highly desired, which will result in lower unit cost.

14.5 Calculating the Total Machining Cost

Previous research in the field of production engineering has shown that the total machining cost is the sum total of the machine cost, setup cost, tooling cost, and power consumption cost due to machining.

$$T_{mc} = M_c + S_c + T_c + P_c$$

T_{mc} = Total maching cost, M_c = maching cost, S_c = setup cost,
T_c = tooling cost, and P_c = power consumption cost

Generally speaking, machining cost is a function of the actual capacity utilization of the machine. Although the installed capacity of a machine may actually be 24 h per day, the available capacity is predetermined by the number of hours it is worked in a day. However, the actual capacity is usually much lower, as a result of downtime, when the machine is not in productive use. Downtime encompasses production time loss because of tool change, setup, waiting, maintenance both scheduled and unscheduled, as well as workers' recess or break time. Research by Ratnasingam (2015) has shown that generally capacity utilization for most machines in furniture factories is in the range of 90%, and factory layout and materials flow on the factory shopfloor play a significant role in production time loss.

14.6 Total Manufacturing Cost of Furniture

The total manufacturing cost of furniture is the sum total of direct materials, direct labor, total indirect costs, and the accrued period costs. Figure 14.3 illustrates the cost distribution in furniture manufacturing that is commonly practiced (Ratnasingam 2015).

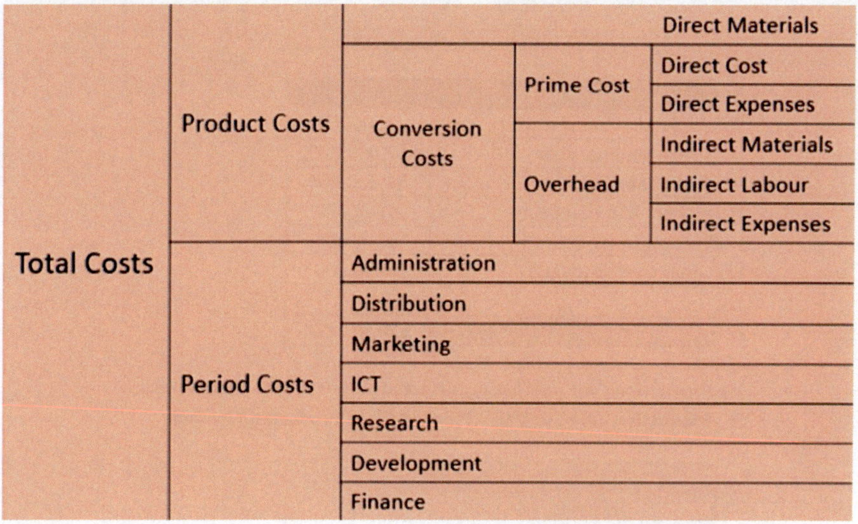

Fig. 14.3 Cost distribution in furniture

Direct materials costs include the cost of all materials that can be accurately apportioned to the product, while *indirect materials costs* are materials that are consumed with no accurate apportionment to a particular product. Examples of direct materials are wood, wood-based panels, adhesive/glue, finishing, fittings, and carton box (Sheu et al. 2003). On the other hand, indirect materials represent consumables, i.e., sandpaper, packaging material, adhesive tape, etc.

Likewise, *direct labor costs* that all labor cost that can be attributed to the product, while indirect labor cost is costs associated with supervisory and production planning. Example of direct labor is the production workers on the factory shopfloor, working on a particular product.

Period costs will cover all other costs incurred to produce all the products in the factory. Much of this cost arises from supporting functions, such as administrative, finance, marketing, operations control, security, human resource management, safety and health, and research and development (Omachonu et al. 2004).

In furniture production however, a batch production often involves a sizeable volume, and therefore, the unit cost of the product is derived by averaging out all direct, indirect, and overheads cost incurred against the total volume produced (Anon 2015) (Fig. 14.4).

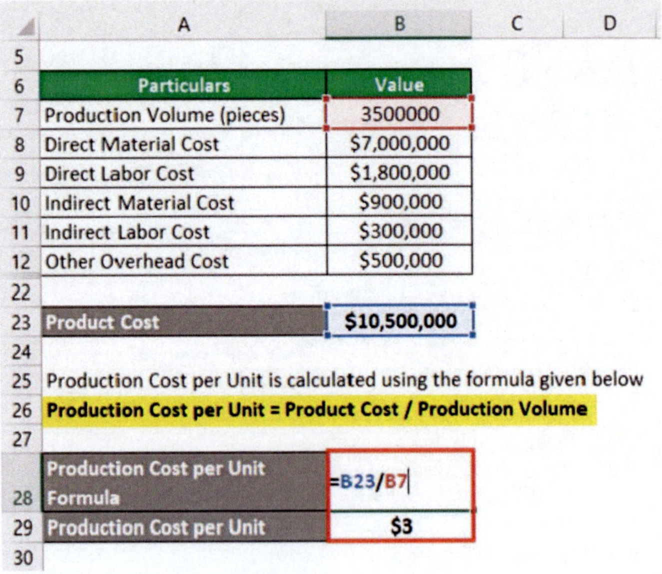

	A	B	C	D
5				
6	Particulars	Value		
7	Production Volume (pieces)	3500000		
8	Direct Material Cost	$7,000,000		
9	Direct Labor Cost	$1,800,000		
10	Indirect Material Cost	$900,000		
11	Indirect Labor Cost	$300,000		
12	Other Overhead Cost	$500,000		
22				
23	Product Cost	$10,500,000		
24				
25	Production Cost per Unit is calculated using the formula given below			
26	**Production Cost per Unit = Product Cost / Production Volume**			
27				
28	Production Cost per Unit Formula	=B23/B7		
29	Production Cost per Unit	$3		
30				

Fig. 14.4 Costing sheet

14.7 Standard Costing

One of the major challenges in furniture manufacturing is the inability of manufacturers, especially the small and medium sized enterprises (SMEs) to accurately calculate their raw material costs used in the product. The mathematical formula provided in the previous sections is aimed at assisting them to establish a reliably, accurate quantities of the different raw materials used, subject to the agreed allowable allowances in the respective factory. In this respect, the concept of standard costing is not only pertinent but is also a very relevant tool for furniture manufacturers to establish the raw materials cost. To ensure that raw materials utilization is well managed in furniture manufacturing, the standard costing method is increasingly used in the furniture industry (Ratnasingam 2015).

Standard cost is an important subtopic of cost accounting. Standard costs have been associated with a manufacturing company's costs of direct materials, direct labor, and manufacturing overhead. Generally, furniture manufacturers would assign the costs of direct materials, direct labor, and manufacturing overhead to a product, rather than the actual costs. This means that a manufacturer's inventories and cost of goods sold will begin with amounts that reflect the standard costs, not the actual costs, of a product. Since a manufacturer must pay its suppliers and employees the actual costs, there are almost always differences between the actual costs and the standard costs, and the differences are noted as *variances* (Ratnasingam 2015).

These variances tell the factory management that the actual manufacturing costs are different from the standard costs. Management can then direct its attention to the cause of the differences from the planned amounts. The variance analysis breaks down the variation between actual cost and standard costs into various components (volume variation, material cost variation, labor cost variation, etc.). This is an important management tool as it enables managers to understand *why costs were different from what was planned* and take appropriate action to correct the situation. In many furniture factories, the *cost-volume-profit analysis* is also practiced, examining the relationship between selling prices, sales, production volumes, costs, expenses, and profits. This analysis provides very useful information for decision-making in the management of a company. For instance, this analysis helps to establish sales prices, in the product mix selection to sell, in the decision to choose marketing strategies, and in the analysis of the impact on profits by changes in costs. In the globalized and highly competitive furniture market, managers must act quickly based on sound scientific and costing information.

The relationship between the cost, volume, and profit is known as the *contribution margin*. The contribution margin is the revenue excess from sales over variable costs. The concept of contribution margin is particularly useful in the planning of business because it gives an insight into the potential profits that a product-line or product mix can generate.

In a survey of furniture manufacturing SMEs in the Southeast Asian region (Ratnasingam 2015), it was shown that standard costing technique is the most widely adopted cost accounting tool employed within the industry. Its benefits include 1. providing manufacturers with cost data based on carefully predetermined amounts, 2. facilitating the planning for raw materials, labor, and overhead requirements, 3. measuring the expected level of performance on the factory shopfloor, and 4. benchmarking for measuring overall factory's productivity.

14.8 Production Planning and Control

In furniture manufacturing, one key function that often poses a challenge is the production planning and control (PPC) function. PPC is a predetermined process that plans, manages, and controls the allocation of human resource, raw material, and machinery to achieve maximum efficiency (Fig. 14.5). The main objectives of the PPC function are to:

- Ensure cost-efficient production process,
- Promote timely delivery of goods,
- Minimize production time,
- Improve customer satisfaction,
- Coordinate with departments about production, to ensure things are on the same page,
- Ensure the right man is assigned the right work,

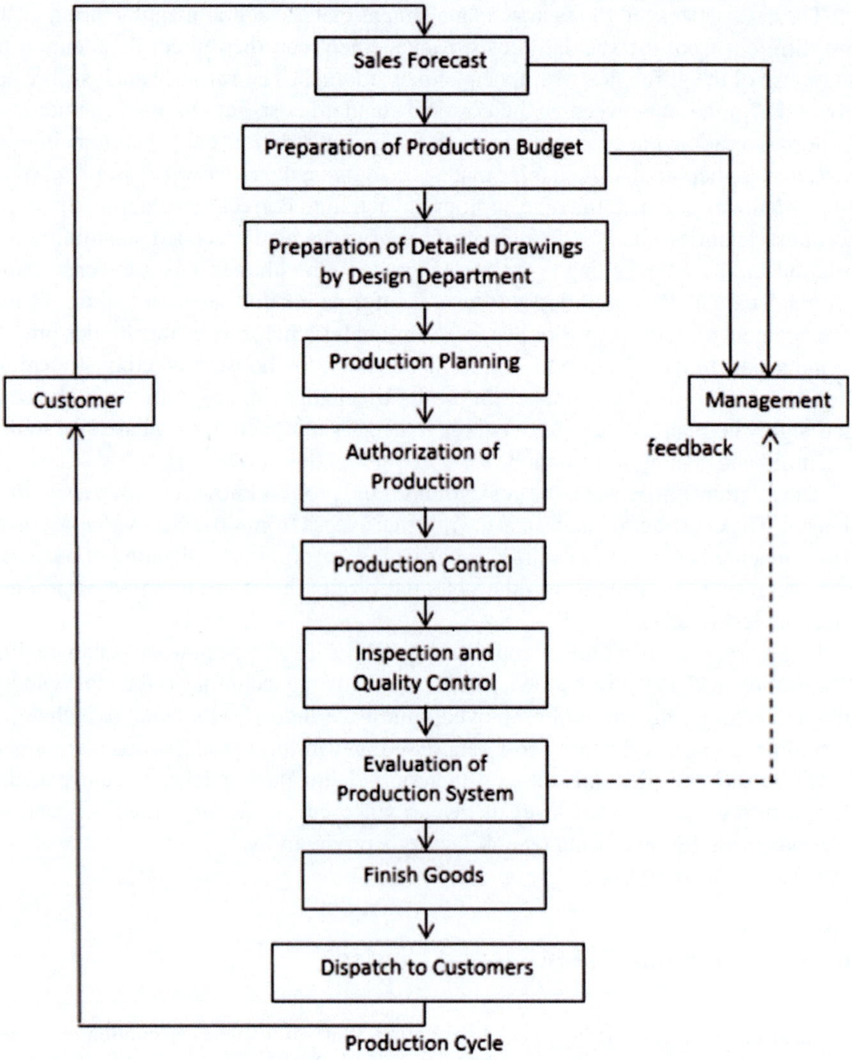

Fig. 14.5 Production planning and control function

PPC is the core of any manufacturing unit. It combines two strategies: production planning and production control. In the manufacturing world, production planning and control are defined by four stages: *Routing*, *Scheduling*, *Dispatching*, and *Follow-Up* (Sheu et al. 2003). The first two stages relate to production planning, while the second two relate to production control.

1. **Routing**
   Routing determines the path from which the raw materials flow within the factory. Once, the sequence is followed, raw materials are transformed into

finished goods. Setting up time for every workstation/machine is important to measure the overall duration of the production process. In essence, routing in manufacturing shows the sequence of work and operations. Routing throws light on the quantity and quality of materials to be used, resources involved (men, machine, and material), the series of operations, and place of production. Routing manages 'How', 'What', 'How much', and 'Where' to produce on the factory shopfloor. *It systematizes the process and nurtures optimum utilization of resources to get the best results.*

To establish the production time at each workstation or machine center, the techniques of method study and time study are used. These techniques constitute work study, which is a powerful tool for production control and planning as it is the scientific basis on which production time is established. It must be emphasized that production time is made up of process time, setup time, handling time, waiting time, tool change time, and downtime. In essence, these various time elements must be accurately measured with the necessary allowances, to establish a standard time, which is both reliable and pragmatic (Anon 2015). As highlighted in the previous sections, the application of mathematical calculations to establish the materials and machining costs will contribute towards this cause, subject to further adjustments to suit the work conditions on the shopfloor. Such an approach will pave the way for greater adoption of production engineering principles in the manufacture of furniture, with greater accuracy and consistency.

2. **Scheduling**
   Scheduling is the second step that emphasizes on 'When' the operation will be completed. It aims to make the most of the time given for completion of the operation. Scheduling is defined as '*the determination of the time that should be required to perform the entire series as routed, making allowance for all factors concerned*'. Manufacturers often use different types of schedules to manage the time element. These include master schedule, operation schedule, and daily schedule.
3. **Dispatching**
   The third step ensures that operations are done successfully, and everything is loaded on the software. *Dispatching includes the release of orders, in accordance with the scheduled charts.*

   The activity of dispatching entails: (1) issue of materials or fixtures that are important for the production usually based on the Bill of Materials (BOM) for the product, (2) issue of orders or drawings for initiating the work, (3) maintain the records from start to end, (4) initiate the control procedure, and (5) cascade the work from one process to another.
4. **Follow-up**
   Follow up, which is also known as expediting, is the final step that finds faults or defects, bottlenecks, and loopholes in the entire production process. *In this step, the team measures the actual performance from start till the end and then compares it with the expected one.* Expediters or progress chasers are responsible for performing follow-up process. It is obvious that any of the

processes may undergo breakdowns or machine failure. Follow-up promotes smooth production by eliminating these defects.

The PPC function relies heavily on time and capacity management to enhance the efficiency of the manufacturing shopfloor. These factors are directly associated with the formulation of effective scheduling of the manufacturing processes (Fig. 14.6). In furniture manufacturing, two types of scheduling are usually done: (1) *Master Production Scheduling (MPS),* and (2) *Manufacturing and Operation Scheduling (MOS)* (Anon 2015).

Master Production Scheduling (MPS) is a scheduling strategy that dictates when and how much of each product is going to be produced based on criteria such as demand, capacity, and inventory availability (Abu et al. 2019). This type of scheduling focuses on a planning horizon that is divided into equal periods (called

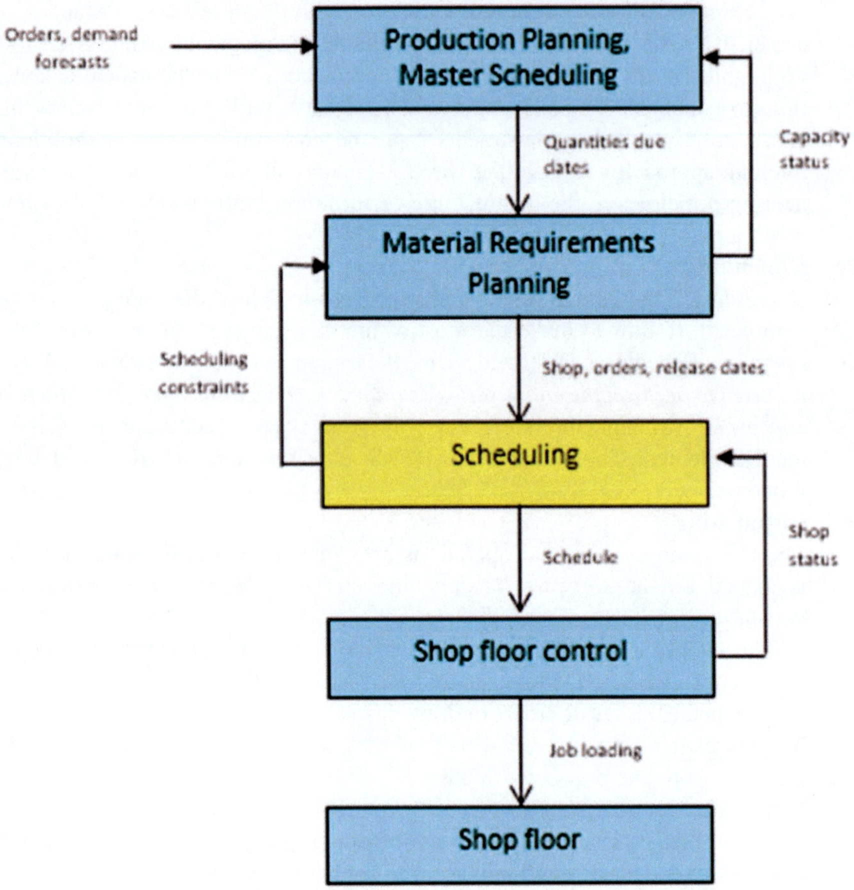

Fig. 14.6 Scheduling function

'time buckets'). It includes a plan to produce certain products and defines resources, staffing, inventory, etc. required for the allotted period.

To assist in the decision-making, MPS will require inputs, such as forecasted demand, production costs, inventory costs, customer needs, production lead time, and manufacturing capacity. The MPS in turn helps to determine the amounts to produce, staff/manpower requirements, quantity of products available for sales, and projected available funds for production. MPS also sets the expectations of the revenue that the business is likely to generate. These outputs can then be used to create a Material Requirements Planning (MRP) schedule.

Manufacturing Scheduling (also called 'Detailed Scheduling' or 'Production Scheduling') focuses on a shorter horizon than MPS. This type of scheduling fixes a time and a date to each operation in a continuous timeline rather than in time buckets (Rewers et al. 2016). Each process can then be visualized in terms of its start time and completion timeframe. The subsequent stages of production planning and control depend on this timeline.

Scheduling aims to optimize the use of time in each step of the production process, from raw or intermediate materials to the delivery of the finished good to the customer. The primary goal is to maximize throughput (output) and on-time delivery within the constraints of equipment, labor, storage, and inventory capacity. Master scheduling and production scheduling work in concert to create capacity and inventory plans that maximize a business's resources to serve its customers efficiently.

Proper use of scheduling methods results in enormous benefits for the furniture manufacturing business. Efficient planning and disciplined schedule execution on the shopfloor can have impacts of up to 25% or higher on output and on-time delivery. Failure to systematize this process can result in lost customers due to late delivery and excessive lead times. In addition, cash flow problems can arise due to delayed shipments, and stressful operating environments are created due to the chaotic and reactive management of resources and orders.

For an efficient and effective PPC function, accurate and timely data from the shopfloor is important. In this context, furniture manufacturers throughout the world are increasingly relying on software such as materials requirement planning (MRP) and enterprises requirement planning (ERP) (Wouters and Stecher 2017). Such planning and scheduling software is becoming a common application in many modern-day manufacturing environments, as customer demand for increased product assortment, fast delivery, and downward cost pressures become prevalent. These systems help planners save time while providing greater agility in updating ever-changing priorities, production schedules, and inventory plans (Fig. 14.7).

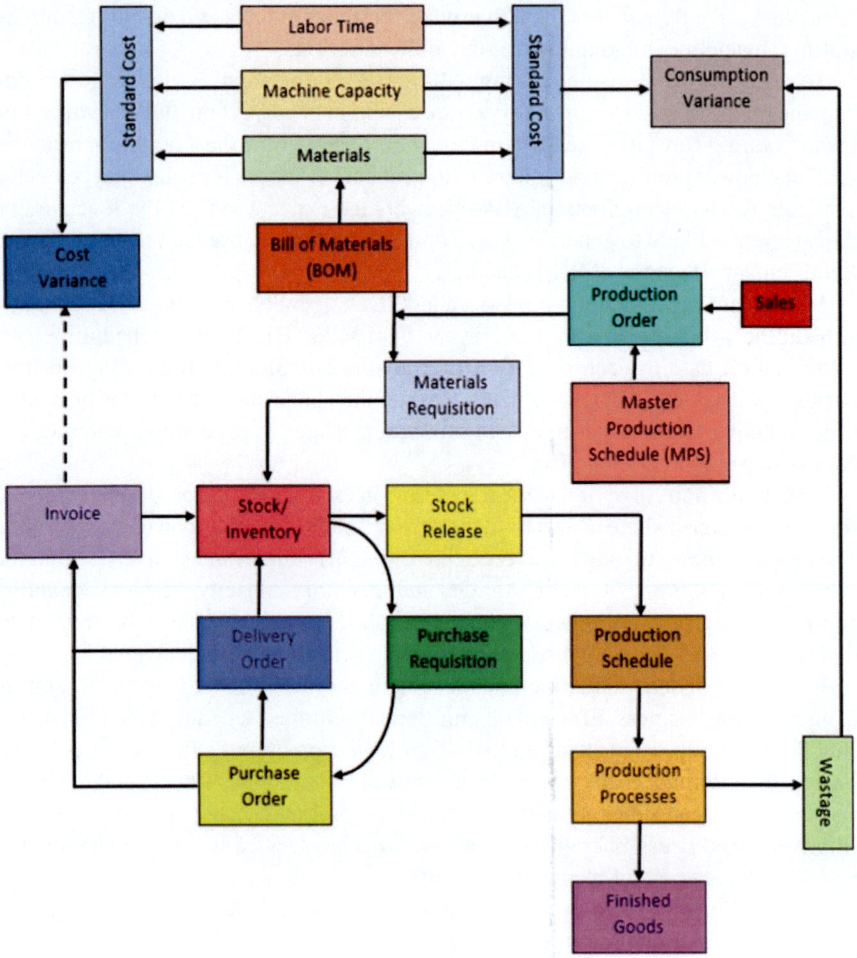

Fig. 14.7 Flow of information in production planning and control

14.9 Work Study

To perform the PPC function successfully, data on time and methods used in the furniture manufacturing factory must be available. This information is often captured through work study techniques. Work study is defined as the analysis of a job for the purpose of finding the preferred method of doing it and determining the standard time to perform it by the preferred (or given) method. It comprises two areas of study: **method study** (motion study) and **time study** (work measurement).

Method study or motion study is mostly used to improve the method of doing work. It aims for the optimum use of best materials and appropriate manpower so that work is performed in well-organized manner leading to increased resource utilization, better quality, and lower costs. It is a technique of analyzing the body motions employed in doing a task to eliminate or reduce ineffective movements and facilitates effective movements. It aims for motion economy by redesigning the task to achieve greater efficiency (Ratnasingam 1999).

Time study or work measurement provides the standard time, that is the time needed by worker to complete a job by the standard method. Standard times for different jobs are necessary for the proper estimation of

- manpower, machinery, and equipment requirements.
- daily, weekly, or monthly requirement of materials.
- production cost per unit as an input to better make or buy decision.
- labor budgets.
- worker's efficiency and make incentive wage payments.

Through the application of method study and time study in the furniture manufacturing industry, it is possible to achieve greater output at less cost and of better quality, and hence achieve higher productivity. It must also be emphasized that work study and ergonomics are two areas of study with the same objective, i.e., designing the work system to be safe for the worker, apart from being productive.

Time study or work measurement refers to the estimation of standard time for a task, i.e., that is the time allowed for completing one piece of job by using the prescribed method. Standard time can be defined as the time taken by an average experienced worker for the job with provisions for delays beyond the worker's control. Standard times for manufacturing processes are important inputs to determine: (1) material, machinery, and manpower requirements, (2) production cost per unit as an input for the preparation of budgets and selling price, (3) manufacturing cycle time and delivery schedule, and (4) workers' performance for payment of incentives (Ratnasingam 2015).

The following steps should be followed to implement a proper time study:

Step 1: Define the task to be studied, and verify the standard method used in the process. Ensure the worker is properly trained and skilled for the task. Select the worker to be studied.

Step 2: Divide the task into reasonably small elements and record them on the time study observation sheet (Fig. 14.8). Time the worker carrying out each of the elements over several cycles.

Step 3: Collect and record the data of required number of cycles by timing and rating the worker. The number of cycles required is often determined statistically depending on the accuracy and confidence level expected of the data.

Step 4: Calculate the representative watch time for each element of operation. Multiply it by the rating factor to get normal time.

$$\text{Normal time} = \text{Observed time} \times \text{Rating factor}$$

Time Study Sheet		
Sheet No. 01	Machine: HSR-04	Edge Routing
Operation: Routing	Attachment/ Guide: N/A	
Material: Oak	Order: F004/2021	
Product: La Redoutte Side Table	ETD: Wk 28	

	Operation Element Breakdown										Foreign elements		
	Pick Up		Place		Machine		Pick Up		Place and Reload				Description
Operation Cycles	T	R	T	R	T	R	T	R	T	R	T	R	
1	5	5	4	9	3	12	9	21	6	27			
2	4	31	7	38	3	41	8	49	0	49			
3	6	55	3	58	2	60	11	71	5	76			
4	6	82	4	86	3	89	6	95	0	95			
5	7	102	3	105	2	107	3	110	0	110			

Summary	Pick Up	Place	Machine	Pick Up	Place and Reload
Total Time	28	21	13	37	11
Abnormal Time	0	0	0	0	11
No of Cycle	5	5	5	5	5
Avg. Time	5.6	4.2	2.6	7.4	0
Rating	100%	100%	100%	100%	100%
Normal Time	5.6	4.2	2.6	7.4	0
Allowances (%)	20%	29%	29%	20%	0
Standard Time (Element)	6.72	5.418	3.354	8.88	0
Standard Time (in Sec.)	24.372				
Standard Time (in Min)	0.4062				

Allowances
Machines Allowances
Tool Change: 9%
No Parts: 5%
Breakdown: 3%
Recess: 10%
Fatigue: 5%

Note: T= Time, R= Clock Reading. Allowances added here are examples only.

Fig. 14.8 Time study sheet

The normal time for the whole operation is then established by adding the normal time of its various elements.

However, it must be recognized that rating factor is a subjective method of applying some adjustment to the mean observed time to arrive at the time that the normal worker would have taken to do that job when working at an average pace. It is based entirely on the experience, training, and judgment of the work-study engineer.

$$\text{Rating Factor} = \text{Observed Performance/Normal Performance}$$

Step 5: Provide allowances for fatigue and various delays.

Step 6: Establish the standard time of manufacturing process.

$$\text{Standard time} = \text{Normal time} + \text{Allowances}$$

Occasionally, some furniture factories may opt to carry out work sampling to establish the real-time variance observed on the factory shopfloor by randomly sampling the time taken for each of the manufacturing process (Xiong et al. 2017). However, work sampling requires the services of skilled work study or industrial engineers, to ensure a proper work sampling is done.

Fig. 14.9 Pricing strategy for furniture

14.10 Pricing Strategy for Furniture

The price of a piece of furniture sold in the market is dependent on several factors, which determines the price-point of the product, beyond which it becomes non-marketable (Fig. 14.9). Furniture being a merchandise and fashion is sold based on *perceived value,* and not on actual value. In this context, most furniture manufacturers price their products based on acceptable price-point of the product, and the prevailing market conditions, while keeping in mind any fluctuation or market trends, to maximize the profit (Wouters and Stecher 2017).

It must be recognized that, for a good pricing strategy for furniture, it is important for furniture manufacturers to appreciate and embrace the concept of *furniture as fashion* and *furniture as lifestyle.* This will inevitably encourage the furniture manufacturers to embark on greater value-added manufacturing of innovative and creative furniture, which in turn will encourage customers to be *willing to pay* (WTP) more for the product (Ratnasingam 2015).

14.11 Lean Manufacturing

Perhaps the most popular method used for cost optimization practices in the furniture manufacturing industry is *lean manufacturing* (LM). Lean manufacturing is a production method that was adapted from Toyota's 1930 operating model, known as the *Toyota Way*. The concept of LM is simply about specifying value by specific product, identifying the value stream involved, ensuring value flow without interruptions, and allowing the customers to draw value from the manufacturer. In essence, lean manufacturing is focused on pursuing manufacturing perfection (Ikumapayi et al. 2020).

From another perspective, lean manufacturing can be described as a method of doing more, with less (Abu et al. 2019). It involves less human effort, less equipment, less time, and less space, while striving to provide the customer's expected product (Fig. 14.10). The lean manufacturing method involves five key principles:

(1) *Value*—Specifying the customer's preference.
(2) *Value Stream*—Identifying the value stream for the product, with no waste.
(3) *Flow*—Making the value-adding strep continuously without interruptions.
(4) *Pull*—Linking all synergistic steps to facilitate continuous flow.
(5) *Perfection*—Managing toward perfection to ensure production proceeds with minimal cost.

Lean manufacturing is founded on the concept of continuous and incremental improvements on processes, and product, while eliminating redundant activities. Only activities that add value that the customer is willing to pay for are considered, while everything else is waste and should be eliminated, simplified, reduced, or integrated. Research has shown that lean manufacturing is a systematic improvement of manufacturing operation, which serves as a precursor to the adoption of automation and technology (Rewers et al. 2016).

A supporting tool of lean manufacturing is the *5S method*, which was also pioneered in Japan. It is a workplace organization method, which emphasizes on the effectiveness and efficiency of the workplace, involving principles such as *sort, set in order*, *shine*, *standardize*, and *sustain.* The concept of 5S is frequently taken as a construct of visual control, visual workplace, or visual factory (Chandrayan et al. 2019). In essence, 5S describes how to identify and store items being used, maintain

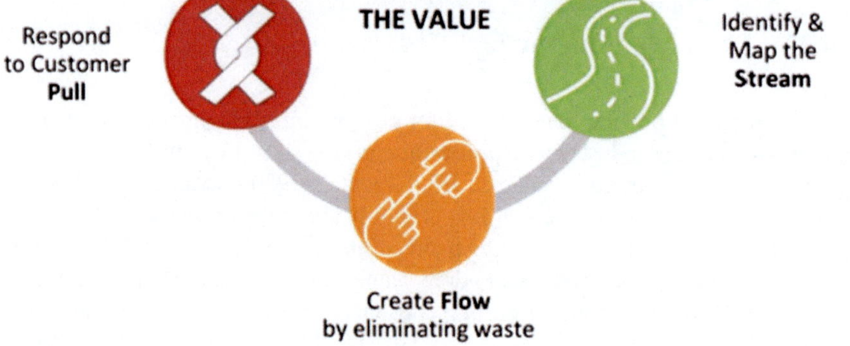

Fig. 14.10 Principles of lean manufacturing

Fig. 14.11 The 5S method

the area and the items, as well as sustaining the new order of the workplace to create a high productivity workplace (Fig. 14.11). In this context, the decision-making is always through dialogue, consensus, and standardization, which in turn help build a high level of understanding and trust among employees on how best to carry out their work (Randhawa and Ahuja 2017).

In the final analysis, it is apparent that cost optimization in furniture manufacturing is the pre-requisite for successful management of the manufacturing operation. In many furniture manufacturing operations, the limited applications of engineering and mathematical approaches hinder the successful application of optimization principles, which inevitably compromises the profitability and competitiveness of the business (Xiong et al. 2017).

Summary

- The application of engineering and mathematical principles to furniture manufacturing enables accurate assessment of the various elements of manufacturing costs.
- Cost optimizations in furniture ensure a more viable manufacturing environment, which also helps improve business profitability.
- Management tools such as lean manufacturing and the 5S method improve the overall manufacturing processes.

References

Abu F, Gholami H, Mat Saman MZ et al (2019) The implementation of lean manufacturing in the furniture industry: a review and analysis on the motives, barriers, challenges, and the applications. J Clean Prod 234:660–680

Anon (2015) Improve your business—costing. International Labor Organization (ILO) Publication, Geneva, Switzerland

Chandrayan B, Solanki AK, Sharma R (2019) Study of 5S lean technique: a review. Int J Prod Qual Manag 26(4):469–491

Ikumapayi OM, Akinlabi ET, Mwema FM et al (2020) Six sigma versus lean manufacturing—an overview. Mater Today-Proc 26(2):3275–3281

Omachonu VK, Suthummanon S, Einspruch NG (2004) The relationship between quality and quality cost for a manufacturing company. Int J Qual Reliab Manag 21(3):277–290

Randhawa JS, Ahuja IS (2017) 5S—a quality improvement tool for sustainable performance: literature review and directions. Int J Qual Reliab Manag 34(3):334–361

Ratnasingam J (1999) Furniture costing in perspective. Sysdata Network Publication, Kuala Lumpur, Malaysia

Ratnasingam J (2015) Mathematics for the furniture industry—managing productivity, costs and quality through calculations. Tech. Note. No. 5, IFRG Publication, Singapore

Rewers P, Trojanowska J, Chabowski P (2016) Tools and methods of lean manufacturing—a literature review. In: Proceedings of 7th international technical conference technological forum, 28–30 June 2016, Czech Republic, pp 135–139

Sheu C, Chen M, Kovar S (2003) Integrating ABC and TOC for better manufacturing decision making. Integr Manufactur Syst 14(5):433–441

Wouters M, Stecher J (2017) Development of real-time product cost measurement: a case study in a medium-sized manufacturing company. Int J Prod Econ 183(Part A): 235–244

Xiong X, Guo WJ, Fang L et al (2017) Current state and development trend of Chinese furniture industry. J Wood Sci 63:433–444

Chapter 15
Emerging Trends in the Global Furniture Industry

Furniture manufacturing is a rapidly growing global industry, which is closely linked not only to demographics, but also the economic performance, as well as global trends impacting the global population. The transiting manufacturing strategies employed leads to a gradual development of the industry in terms of the extent of value-addition. On the other hand, the effect of climate change, circular economy, and the rapid pace of development in the digital economy is also having a profound impact on the way business is conducted in the furniture sector. This chapter describes the main trends shaping the furniture manufacturing industry globally.

15.1 Introduction

Furniture design and manufacturing is an evolving industry, in which the manufacturers strive to gradually move up the value-chain, through a more effective management of the supply chain (Florio et al. 1998; Drayse 2008, 2011). The manufacturing strategies used by furniture manufacturers often begin with being a contract manufacturer, and as they gain greater innovative and creative capabilities, they move up one notch to become a manufacturer of designer furniture, and finally reach the pinnacle of the industry by establishing themselves as a manufacturer of branded furniture (Fig. 15.1).

An *original equipment manufacturer (OEM)* is a company that is contracted by another manufacturer, or a retailer, to produce specific products. In the case of furniture manufacturing, the concept of OEM is prevalent among large manufacturers, who often work with many sub-contractors, each producing parts/components that make up the final furniture (Han et al. 2009). It is merely focused on manufacturing the product, with clear cut instructions from the buyer. In reality, OEM paves the way for out-sourcing activities, which reduces capital expenditure, and also allows economies of scale in production to keep production cost low.

J. Ratnasingam, *Furniture Manufacturing*, Design Science and Innovation,
https://doi.org/10.1007/978-981-16-9412-7_15

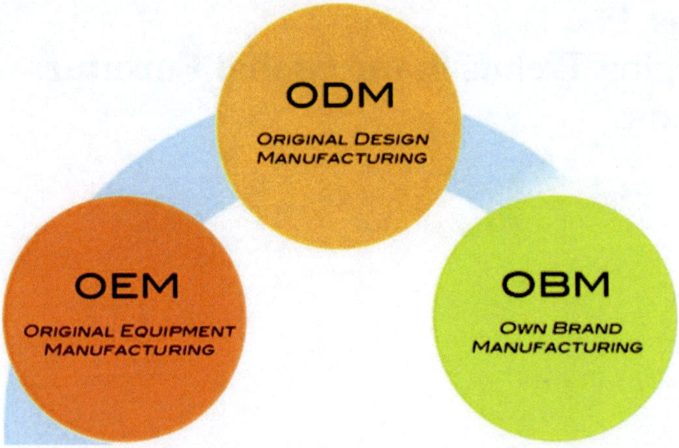

Fig. 15.1 Manufacturing strategies

An *original design manufacturer (ODM)* is a company that designs and manufactures a specific product, which may be sold by another company under its brand. In essence, the company often has several designs of its own, which is sourced and sold exclusively by the buyer, under its own brand. In some instances, the existing designs are slightly modified to suit the buyer's preference. Unlike OEM manufacturers who supply many companies, ODM manufacturers are design capable and are often legally bound to sell the product to a particular buyer, who then sells it under an own brand. The ODM model is gaining popularity in the globalized international trade, where a local ODM manufacturer is licensed to produce goods for a foreign company, that draw some benefits from such engagement, through lower labor cost, better logistics, or proximity to the market (Magistretti et al. 2019). This model is also used to circumvent trade restrictions, where local ownership-laws possibly prohibit direct ownership of assets by foreigners, allowing a local firm to produce for a brand company, either for the domestic market or for export.

An *own/original brand manufacturer* (OBM) is typically a company that is involved in the entire supply chain of the branded product, which may be made by one or more companies, or may include components produced by others (Scott 2006; Ratnasingam 2019). Branding adds high value to a product, and adding a brand signet to an anonymous product often boost its perceived value. The value of the product is raised with branding, as it imparts prestigiousness, assurance of qualities relevant to the user, reliability, and improving the customer's pride of ownership. In this context, furniture designers also lend their names to the product, which inevitably creates a brand. Specialty furniture stores, also sell products under their brand, which provides a competitive advantage, when compared to the unbranded products of similar quality (Fig. 15.2).

As illustrated in Fig. 15.3, the different manufacturing strategies in furniture manufacturing contribute value to furniture through different activities, whether it is pricing, design, or brand perspective. Nevertheless, it must be emphasized that

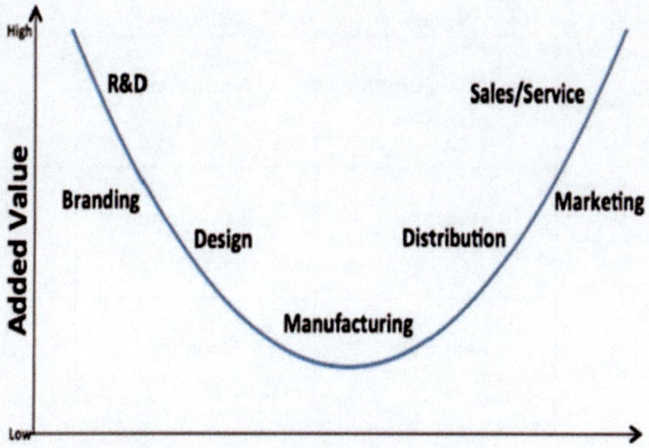

Fig. 15.2 Added-value activities in the furniture industry

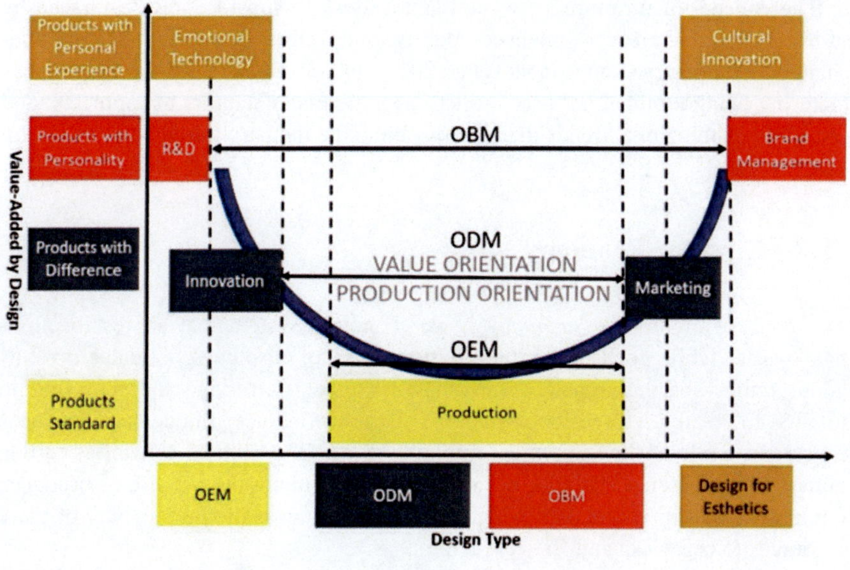

Fig. 15.3 Value addition in furniture manufacturing

the extent to which value is added to furniture is often a function of the product's perceived value, a characteristic determined by the customer, not by the manufacturer.

The fact that furniture is fashion, and it is purchased based on perceived value, rather than actual value must be fully appreciated. Hence, any initiative, such as branding, which will shape and improve the perceived value of furniture, is a worthy effort that is often exploited by furniture manufacturers, and retailers (Ratnasingam 2019). Brand creates a difference in the crowded marketplace, and a brand signet and the related brand advertising are marketing efforts often pursued by retailers to trigger purchase intention among the potential customers (Fig. 15.4).

	Needs	Wants	Desires
User Perception	Practical Function *Quality*	Aesthetics *Qualia*	Symbol *Brand*
Designer Perception	Visceral *Function*	Behavioral *Design*	Reflective *Feeling*
Maker Perception	Feasibility *OEM*	Viability *ODM*	Desirability *OBM*

Fig. 15.4 Manufacturing strategies and consumer perception

The concept of furniture as fashion and a lifestyle, must be fully embraced by furniture manufacturers to remain competitive and facilitate their move up the value-chain of the industry (Ratnasingam et al. 2018). In a similar manner, furniture manufacturing being a global industry, necessitates the manufacturers to appreciate and comprehend the global trends that are impacting the industry throughout the world.

15.2 Circular Economy

The concept of *circular economy* is an economic system that strives to eliminate waste, while focused on the continual use of resources. Circular systems adopt reuse, sharing, repair, refurbishment, remanufacturing, and recycling in totality, to create a closed-loop system. This in turn reduces the demand for input resources, while at the same time, minimizing waste, pollution, as well as carbon emissions. In essence, circular economy aims to prolong the use life of products, equipment, and infrastructure, which inevitably enhances the productivity of these resources (Koszewska and Bielecki 2020).

In the circular economy, waste from one process should become the feed-stock for another, usually as a by-product, or in the form of recovered resource for another industrial process. In other instances, it is converted into a regenerative resource for nature, in the form of compost. Unlike the traditional linear economy, which adopts the production concept of take, make, and dispose, the regenerative approach is embedded in the realms of sustainability (Fig. 15.5).

The furniture industry is a likely candidate for the widespread adoption of the concept of circular economy. Furniture is passive durable products, and therefore, repairing and remanufacturing would prolong the use life of these products, which inevitably leads to lower environmental impacts as well as lower cost. For instance, the European Union (EU) has recognized a huge potential for implementing the concept of circular economy in the furniture sector. It has been reported that almost

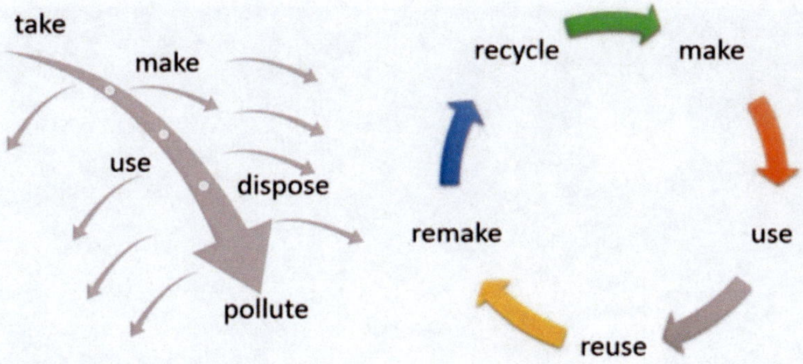

Fig. 15.5 Linear and circular economies

all of the 10,000,000 tons of annually discarded furniture in the EU ends up in landfills or is incinerated. A circular model will in turn give a gross value-added of up to €4.9 billion in gross value-added, while creating 163,300 jobs by 2030. In a study in Denmark, it was found that 44% of the companies had included maintenance in their circular model, while 22% had re-collect schemes, and another 56% paid attention to designing furniture to facilitate recycling (Marica et al. 2019; Susanty et al. 2020). Despite its popularity, circular economy in the furniture industry suffers from a lack of awareness among many companies especially in the developing world, who have limited knowledge on how to make the transition to the circular model. Even in the United Kingdom (UK), the potential for reuse and recycling in the furniture sector has only been recognized over the last decade, and sufficient to say that the pace of things are picking up rapidly, ever since. It has been reported that 42% of the bulk waste sent to landfills annually (1.6 million tons) was furniture, and 80% of the raw material in the production phase is waste. Against this backdrop, it must be emphasized that the concept of circular economy is gaining traction throughout the world, and the furniture sector will have to cope with this emerging trend soon.

15.3 Digital Economy

The concept of *digital economy* refers to an economy that is driven by digital computing technologies, although in another perspective, it is taken as carrying out business in virtual markets based on the Internet and the World-Wide-Web. Therefore, the digital economy is sometimes referred to the *Internet Economy* or *Web-Based Economy* (Sturgeon 2021).

The digital economy involves billions of daily online connections among people, businesses, devices, data, and processes. This extensive interconnectedness is driven through the Internet, mobile technology, and the *Internet of Things* (IoT), which results in the traditional economy being intertwined with the digital economy.

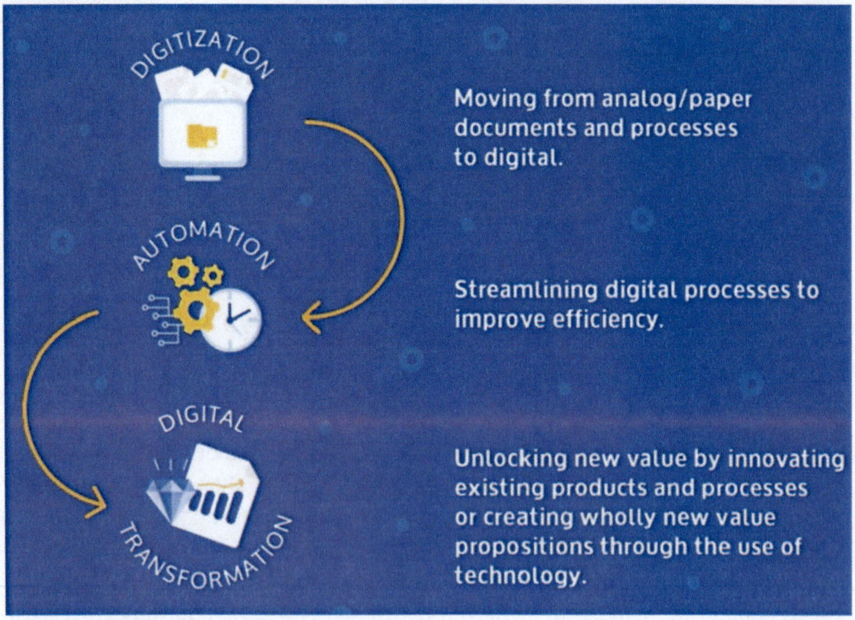

Fig. 15.6 Stages in digital transformation

The digital economy is underpinned by the spread of *information and communication technologies* (ICT) across all economic sectors, enhancing its productivity significantly (Sturgeon 2021). The digital economy is undermining conventional notions about how businesses are structured, how consumers obtain services, information, and goods, and how states need to adapt to these new regulatory challenges. In general, transformation towards the digital economy takes three stages of development (Fig. 15.6).

The first stage is *digitization.* Digitization refers to creating a digital representation of physical objects or attributes. In other words, digitization is about converting non-digital applications into digital representation, which then allows the computerized systems to use it accordingly (Kropivšek and Grošelj 2020). An example from furniture manufacturing would be when the full-size drawing of furniture is transmitted electronically or digitally across the globe from buyers to manufacturers and vice versa. This is the connection between the physical world and software, and it has been done since the1960s. Digitization is an enabler for all the processes that provide business value because of the need for consumable data.

The second stage is *digitalization*, which refers to enabling or improving processes by leveraging digital technologies and digitized data (Petya 2021). Therefore, digitalization presumes digitization. Examples of this from the furniture industry could be as simple as programmable logic control (PLC) in a microprocessor-based system, sequenced logic for a batch process, generating a work order in the enterprise resource planning (ERP) maintenance system, etc. As a matter of fact, digitalization increases productivity and efficiency, while reducing costs. Digitalization improves an existing business process or processes, but doesn't change or transform them. Simply put, it transforms the process from a human-driven event, or series of events to being software driven.

The third stage is *digital transformation*, which in essence is the real business transformation enabled by digitalization (Ratnasingam et al. 2021). The 'digital' moniker is a little bit of a misnomer because the essence of digital transformation is the changing business processes enabled by digitalization technologies. Information technology (IT) and operational technology (OT) have long existed in their separate spheres. Each had its own network, objectives, and requirements. And, until relatively recently, this separation was perfectly fine. However, with the advent of the Internet of Things (IoT), the game is changing rapidly, and tearing down old silos. Complex machines, including robots, are now integrating with network sensors, and being managed by advanced analytics software, blurring the line between where IT ends and OT begins. When IT and OT join forces, organizations unlock several benefits from process automation and business intelligence to a simplified system for ensuring regulatory compliance (Sturgeon 2021). One example of this is the convergence of IT/OT where the intersection and overlap of IT skills within the OT domain has created the need for a more uniform governance, due to cybersecurity concerns, data flow requirements, and skills. Another example of digital transformation is a shift from local control of physical processes to remote monitoring and control of those same processes. An example of digital transformation from IKEA, the global furniture retailer, is the integration of its customer sales volumes with the raw material vendors, hence increasing the efficiency and response of its supply chain (Ratnasingam et al. 2021).

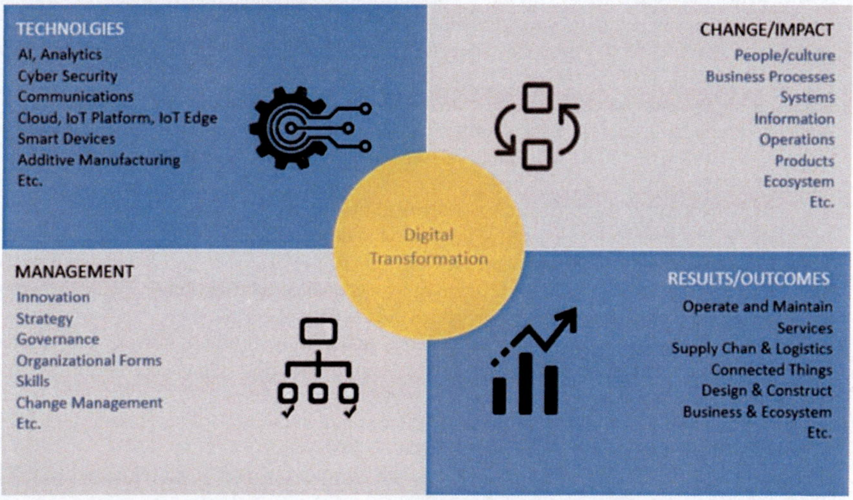

Fig. 15.7 Digital transformation and industry 4.0 (Courtesy of the Malaysian Furniture Council)

From another perspective, it is also apparent that digital transformation is synonymous with the adoption of Industry 4.0, which is becoming increasingly prevalent in many manufacturing industries throughout the world (Fig. 15.7). The concept of Industry 4.0, although somewhat practiced in a slightly different mode in different countries, shares much similarity in which it aims to promote a close interface been IT/OT to enable business growth and sustainability (Ya-Ling and Patarapong 2019).

Summary

- The global furniture industry is an evolving industry subject to market and consumer requirements.
- It is important for furniture manufacturers to closely watch and embrace the evolving market trends to ensure their relevance and competitiveness.
- Circular economy and digital transformation/Industry 4.0 are the two major trends impacting the furniture manufacturing industry throughout the world.

References

Drayse MH (2008) Globalization and regional change in the U.S. furniture industry. Growth Change 39(2): 252–282

Drayse MH (2011) Globalization and innovation in a mature industry: furniture manufacturing in Canada. Reg Stud 45(3):299–318

Florio M, Peracchi F, Sckokai P (1998) Market organization and propagation of shocks: the furniture industry in Germany and Italy. Small Bus Econ 11:169–182

Han X, Wen Y, Kant S (2009) The global competitiveness of the Chinese wooden furniture industry. Forest Policy Econ 11(8):561–569

Koszewska M, Bielecki M (2020) How to make furniture industry more circular? The role of component standardization in ready-to-assemble furniture. Entre Sustain Iss 7(3):1688–1707

Kropivšek J, Grošelj P (2020) Digital development of Slovenian Wood Industry. Drvna Ind 71(2):139–148

Magistretti S, Dell'Era C, De Massis et al (2019) Exploring the relationship between the types of family involvement and collaborative innovation in design-intensive firms: insights from two leading players in the furniture industry. Ind Innov 26(10): 1121–1151

Marica B, Laura B, Elisabetta S (2019) Sustainability and quality management in the Italian luxury furniture sector: a circular economy perspective. Sustain-Basel 11(11):3089

Petya S (2021). Trends in business processes management in the furniture industry in Bulgaria. Sixth international scientific conference 'business and regional development. SHS Web Conf 120: 02012

Ratnasingam J, Chin KA, Latib HA et al (2018) Innovation in the Malaysian Furniture Industry: drivers and challenges. BioResources 13(3):5254–5270

Ratnasingam J (2019) Emerging trends in the global furniture market. Tech. Note, no 8. IFRG Publication, Singapore

Ratnasingam J, Ioras F, Lim CL et al (2021) Digital technology application among malaysian value-added wood products manufacturers. BioResources 16(2):2876–2890

Scott AJ (2006) The changing global geography of low technology, labor-intensive industry: clothing footwear and furniture. World Dev 34(9):1517–1536

Susanty A, Tjahjono B, Sulistyani RE (2020) An investigation into circular economy practices in the traditional wooden furniture industry. Prod Plan Control 31(16):1336–1348

Sturgeon TJ (2021) Upgrading strategies for the digital economy. Glob Strateg J 11:34–57

Ya-Ling H, Patarapong I (2019) Alternative technological learning paths of Taiwanese firms. Asian J Technol Innov 27(3):301–314

Chapter 16
Conclusions

Furniture manufacturing is as old as human civilization, and the demand for furniture around the world has been steadily growing over the years. However, the onset of global Covid-19 pandemic in late 2019 has brought about some major changes and trends that will have a significant impact on the furniture manufacturing industry and its trade in years to come. The following are the ten most important trends that will continue to shape the furniture sector throughout the world, in the years to come (Ross 2016; Schwab 2018; Ratnasingam 2019).

1. *Rising Online/Digital Shift*

   Manufacturers, designers, marketers, and consumers are increasingly adopting e-commerce, as the traditional marketplaces appear to be losing its appeal and also in compliance with the new norm.
2. *Omni Channel Sales*

   To cope with the new norm, traditional marketing activities are being supplemented with digital marketing, to provide customers with multiple channels of sales. Although physical stores are gradually opening, it is almost certain that digital marketing is accelerating in terms of its application in the furniture sector.
3. *Digitalization of Manufacturing*

   The application of digital technologies and technologies of Industry 4.0 will surely pervade throughout the furniture sector, as many traditional manual manufacturing jobs will be replaced by machines and technology sooner than later.
4. *Customization and Modularity*

   Customers are increasingly looking for furniture that are both customized and modular in nature. Inevitably, technology application will not only shape such furniture design, but also its manufacturing.

J. Ratnasingam, *Furniture Manufacturing*, Design Science and Innovation,
https://doi.org/10.1007/978-981-16-9412-7_16

5. *Smart Furniture*

   As homes and dwellings changes in line with the trends in urbanization, the demand for smart furniture will also increase in years to come. Such multi-functional furniture will not only provide comfort, but also serve needs of customers whose living space are reduced.

6. *Remote Office Furniture*

   The onset of the Covid-19 pandemic, has completely transformed the working space, as more and more employees are opting to work from home (WFH). In this context, the demand for small office-home office (SOHO) type of furniture has increased exponentially throughout the world.

7. *Marketing to Millennials*

   As most millennials are tech-savvy, the traditional marketing approach is no longer effective. The application of augmented reality (AR) to give the potential customers a personalized experience, is becoming the norm in marketing furniture. In this respect, furniture marketeers and merchandisers must also be tech-savvy in order to be able to communicate and convince the millennials to purchase the furniture of their preference.

8. *Cash and Carry Option*

   The Covid-19 pandemic has not only affected the global economy severely, but it also reduced earnings and wage levels throughout the world. As disposable income is slashed among consumers, the cash and carry option among furniture buyers is becoming increasingly common throughout the world. In this respect, the traditional built-up furniture is losing its appeal, as more and more manufacturers are shifting towards manufacturing easy to assemble and disassemble furniture, which is often sold on a cash and carry option.

9. *Eco-Friendly Furniture*

   With increasing global environment consciousness and the need to mitigate the climate change phenomenon, consumers are increasingly demanding eco-friendly furniture. Apart from being in line with the Sustainable Development Goals (SDGs) , consumers are also insisting on sustainably sourced raw materials to be used in furniture.

10. *Impeccable Logistics and Supply Chain*

    It is undeniable that furniture manufacturing is a global industry with an extensive supply chain spread throughout the world. This characteristic will shape the competitiveness of furniture manufacturers throughout the world, and those who fail to keep up with the global trend will be left behind. In fact, during the Covid-19 pandemic the disruptions in logistics and supply chain have severely impacted many manufacturing industries, including the furniture sector.

Despite the trends elaborated above, it must be emphasized that the furniture manufacturing industry has a low-entry barrier, and remains characteristically cottage-based in many parts of the world. Predominated by small and medium enterprises (SMEs) as well as many micro-companies, the industry is starved of the necessary investments into manufacturing technologies and operational systems. Hence, in many countries around the world, it continues to resemble a craftsmen-oriented trade, where experience is accorded more importance compared to knowledge.

In this context, the furniture manufacturing industry is arguably devoid of engineering perspectives, which is the necessary ingredient to bring about industrial transformation, leading to higher value-addition and productivity.

In order to do so, the provision of the necessary information and knowledge from an engineering perspective that could be easily understood and adopted by students in undergraduate and graduate programs, researchers, industry practitioners, as well as anyone interested in the furniture manufacturing industry is both timely and warranted. Afterall, furniture manufacturing should be considered a branch of production engineering, and application of engineering and mathematical principles would transform the industry into a greater value-added industry, while increasing its global competitiveness, as well as its long-term sustainability.

References

Ratnasingam J (2019) Emerging trends in the global furniture market. Tech. Note. No. 8, IFRG Publication, Singapore

Ross A (2016) The industries of the future. Simon & Schuster Inc., New York, United States of America

Schwab K (2018) Shaping the fourth industrial revolution. World Economic Forum, Geneva, Switzerland

Chapter 17
Further Information

This book does not claim to give the final word on the issues of furniture manufacturing processes but aims to set the scene for a systematic framework from a production engineering perspective, for readers in the woodworking, furniture, and related fields. For further information, the following sources including relevant websites are recommended.

Alasdair Gilchrist (2016). *Industry 4.0—The Industrial Internet of Things.* A-Press Media, California, USA.

Anco Prak & Thomas Myers. (1981). *Furniture Manufacturing Processes.* North Carolina State University Press, Raleigh, North Carolina, USA.

Andy Rae. (2001). *The Complete Illustrated Guide to Furniture & Cabinet Construction.* The Taunton Press Inc., Newtown, Connecticut, USA.

Anon. (2010). *Woodworking Machines—Back to Basics.* Skills Institute Press LLC, east Petersburg, Pennsylvania, USA.

Anon. (2010). *Wood Handbook—Wood as an Engineering Material.* (Centennial Edition). Forest Products Laboratory, Madison, Wisconsin, USA.

Anon. (2017). *Forestry for a Low Carbon Future—Integrating Forests and Wood Products in Climate Change Strategies.* FAO Forestry Paper No. 177, Food and Agriculture Organization (FAO) of the United Nations, Rome, Italy.

Antonio Pizzi & Kash Mittal. (2011). *Wood Adhesives.* CRC Press, Boca Raton, Florida, USA.

Bruce King. (2017). *The New Carbon Architecture—Building the Cool Climate.* New Society Publishers, Gabriola Island, British Columbia, Canada.

Carl Eckelman. (1988). *Effective Principles of Product Engineering and Strength Design for Furniture Manufacturing.* Society of Manufacturing Engineers Publication, Southfield, Michigan, USA.

J. Ratnasingam, *Furniture Manufacturing*, Design Science and Innovation,
https://doi.org/10.1007/978-981-16-9412-7_17

Diana Twede, Susan Selke, Pascal Kamdem & David Shires. (2005). *Cartons, Crates and Corrugated Board: A Handbook of Paper and Wood Packaging Technology.* Destech Publication Inc., Lancaster, Pennsylvania, USA.

Donald Askeland & Wendelin Wright. (2016). *The Science and Engineering of Materials.* Cangage Learning, Boston, Massachusetts, USA.

Edward Clark, John Ekwall, Thomas Culbreth & Rudolph Willard. (1987). *Furniture Manufacturing Equipment.* North Carolina State University Press, Raleigh, North Carolina, USA.

Ernest Joyce & Alan Peters. (1997). *The Techniques of Furniture Making.* Chrysalis Press Plc., London, United Kingdom.

Etele Csanâdy & Endre Magoss. (2013). *Mechanics of Wood Machining.* 2nd Edition. Springer-Nature Publication, Cham, Switzerland.

Etele Csanâdy, Endre Magoss & LâszlôTolvaj. (2015). *Quality of Machined Wood Surfaces.* Springer-Nature Publication, Cham, Switzerland.

Etele Csanâdy, Zsolt Kovâcs, Endre Magoss & Jegatheswaran Ratnasingam. (2019). *Optimum Design and Manufacture of Wood Products.* Springer-Nature Publication, Cham, Switzerland.

Franco Bulian & Jon Graystone. (2009). *Wood Coatings—Theory and Practice.* Elsevier Publishing, Amsterdam, Netherlands.

Gary Rogowski. (2002). *The Complete Illustrated Guide to Joinery.* The Taunton Press Inc., Newtown, Connecticut, USA.

Jegatheswaran Ratnasingam. (1999). *Furniture Costing in Perspective.* Sysdata Network Publication, Singapore.

Jegatheswaran Ratnasingam & Chiaki Tanaka. (2002). *Wood Machining Processes: A Managerial Perspective.* The Tanabe Foundation Publication, Shimane, Japan.

Jegatheswaran Ratnasingam, Freider Scholz, Erwin Friedle and Adrian Riegel. (2004). *Wood Sanding Processes: An Optimization Perspective.* 2004. The Shimbon Kogyo Foundation Publication, Tokyo, Japan.

Jegatheswaran Ratnasingam. (2015). *The Malaysian Furniture Industry—Unravelling its Growth and Challenges to Innovation.* Universiti Putra Malaysia Press, Serdang, Malaysia.

Jerzy Smadzewski (2015). *Furniture Design.* Springer International Publishing, Cham, Switzerland.

Jim Postell (2012). *Furniture Design.* Wiley & Sons. Inc., Hoboken, New Jersey, USA.

Jorge Prieto & Jűrgen Kiene. (2018). *Wood Coatings—Chemistry and Practice.* Vincentz Network Publishing, Hanover, Germany.

Mark Kirwan (2013). *Handbook of Paper and Paperboard Packaging Technology.* Wiley-Blackwell Publishing, Oxford, United Kingdom.

Mikell Groover (2018). *Automation, Production Systems and Computer-Integrated Manufacturing.* Pearson-Prentice Hall Publication, London, United Kingdom.

Nick Rudkin. (2013). *Machine Woodworking.* Routledge Press, Oxford, United Kingdom.

Peter Koch. (1964). *Wood Machining Processes.* Roland Press Company, New York, USA.

Rubin Shmulsky & David Jones. (2019). *Forest Products and Wood Science.* 7th Edition. John Wiley & Sons Ltd., Hoboken, New Jersey, USA.

Rudolph Willard. (1982). *Furniture Construction.* North Carolina State University Press, Raleigh, North Carolina, USA.

Sandor Nagyszalanczy. (1997). *The Wood Sanding Book—A Guide to Abrasives, Machines and Methods.* The Taunton Press Inc., Newtown, Connecticut, USA.

Stuart Lawson. (2013). *Furniture Design—An Introduction to Development, Materials and Manufacturing.* Laurence King Publishing Ltd., London, United Kingdom.

Walters Nsoh. (2009). *Controlling Illegal Logging: An Assessment of Timber Certification as an Instrument to Ensure Legal and Sustainable Timber Production.* VDM Verlag, Saarbrucken, Germany.

Useful Websites

AKZONOBEL—coatings (www.akzonobel.com).

AKZONOBEL—adhesives (www.woodadhesives.akzonobel.com).

CSIL Milano—Centre for Industrial Studies (www.csilmilano.com).

European Panel Federation (www.europanels.org).

Food and Agricultural Organization of the United Nations (www.fao.org).

Furniture Industry Research Association (www.fira.co.uk).

Furniture Market in High Point, NC (www.imchighpointmarket.com).

Furniture Today (www.furnituretoday.com).

Furniture Testing (www.furnitest.com).

Health and Safety Executive (www.hse.gov.uk).

HERMES—coated abrasives (www.hermes-schleifwerkzeuge.com).

HOMAG—woodworking technology supplier (www.homag.com).

International Tropical Timber Organization (www.itto.int).

International Bamboo and Rattan Organization (www.inbar.int).

International Association of Packaging Research Institutes—packaging (www.iapri.org).

LEITZ—tooling (www.leitz.org).

National Upholstery Association (www.naturalupholstery.com).

SCM Wood—machines and systems for wood processing (www.scmgroup.com).

Timber Research and Development Association (www.trada.co.uk).

United Nations Economic Commission for Europe on Forestry and Timber (www.unece.org/forests).

United Nations Industrial Development Organization (www.unido.org).

Woodworking Community Web (www.woodworkweb.com).

Ratnasingam—Plates and Images

Plate 1: Trade Map of Furniture Exporters and Importers

See Plate 1.

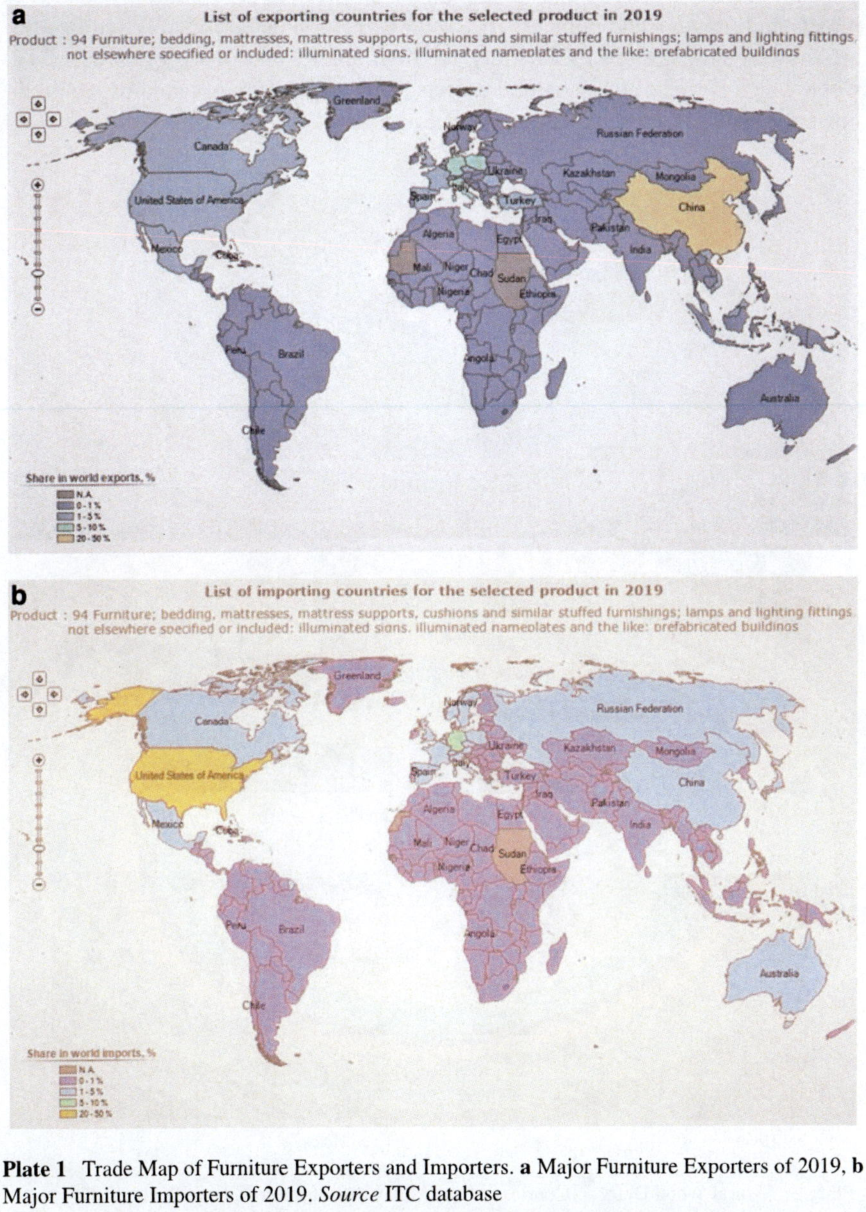

Plate 1 Trade Map of Furniture Exporters and Importers. **a** Major Furniture Exporters of 2019, **b** Major Furniture Importers of 2019. *Source* ITC database

J. Ratnasingam, *Furniture Manufacturing*, Design Science and Innovation,
https://doi.org/10.1007/978-981-16-9412-7

Plate 2: Natural Wood Defects

Wood being a natural material, is subjected in variations not only in its physical properties but also in its chemical and mechanical properties. The variations can be attributed to the inherent features of the wood (i.e., genetic and physiology of growth), but may also arise due to external factors, such as biodegrading agents (i.e., insects and fungus), and climate. The common natural defects manifested in wood are shown below (Plate 2). Generally, the high quality of wood is often defined as wood which is free of these natural defects. The wood grading standards would list down the criteria for each quality grade class.

Pith

Knot

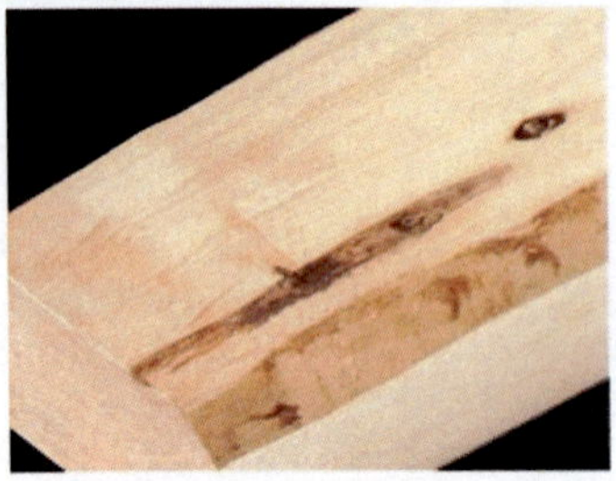

Wane

End Split

Pin Holes

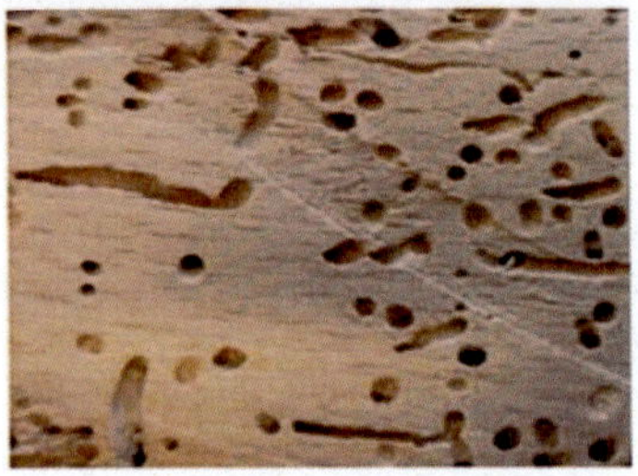

Grub Holes

Plate 2 Natural Wood Defects (Courtesy of the Malaysian Furniture Council)

Worm Infestation

Wood Stain/Blue Stain

Wood Decay/Rot

Wood Grain Variations

Plate 2 (continued)

Plate 3: Wood Drying Defects

Wood being a hygroscopic material, must be subjected to drying or seasoning, to remove the excess moisture present in it, to an acceptable level. This acceptable level is often known as the equilibrium moisture content (EMC), or the point at which the wood remains stable in the environment with a balanced adsorption/desorption of moisture from the surrounding. Generally, wood which has a high moisture content is of lower strength, unstable, and is susceptible to degradation. If the drying/seasoning of the wood takes place in a non-optimal rate, coupled with the natural characteristics of wood, some drying defects may manifest, which may adversely impact the quality of the wood (Plate 3).

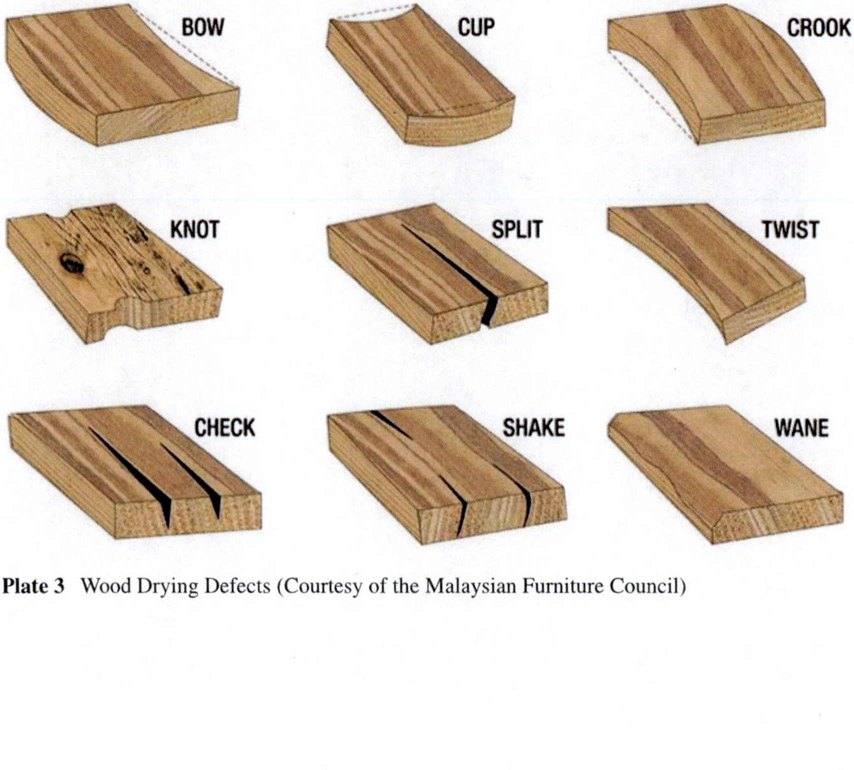

Plate 3 Wood Drying Defects (Courtesy of the Malaysian Furniture Council)

Plate 4: Wood Machining Defects

Machining defects can arise on the wood surface when the machining operations are carried out under non-optimal conditions. This may be due to poor choice of cutting tool and its geometry, improper machining process parameters, and poor quality of wood or workpiece. Further, some natural properties of wood may cause the manifestation of particular machining defects. The common wood machining defects encountered in the wood machining processes are as shown below (Plate 4).

Fuzzy Grain

Tear Out

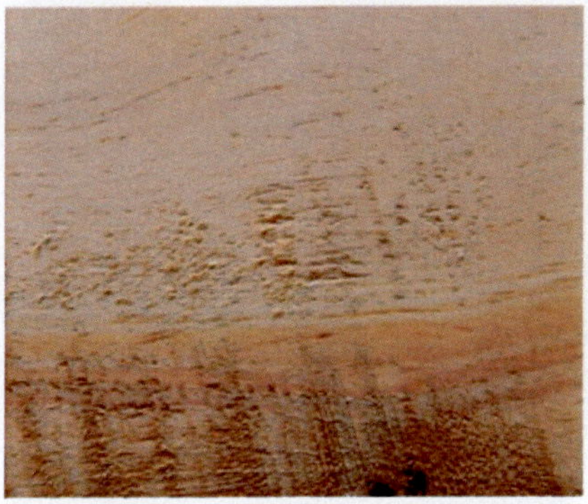

Chip Out

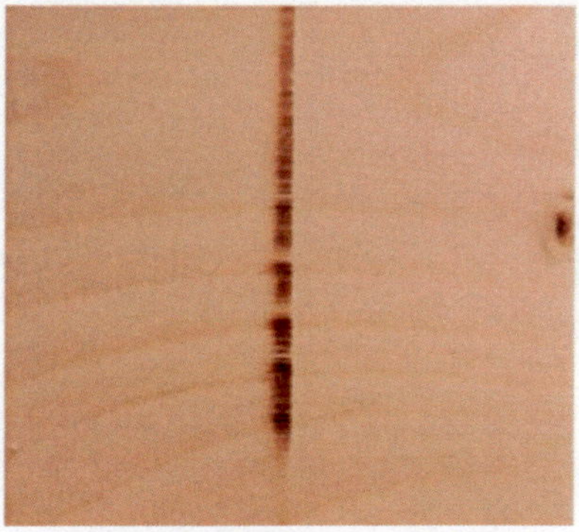

Cutter Burn Mark

Plate 4 Wood machining defects (Courtesy of the Malaysian Furniture Council)

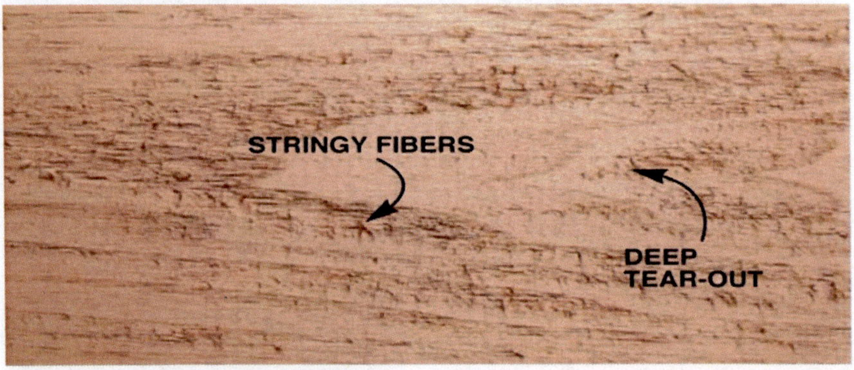

Stringy Fibers and Deep Tear Out

Raised Grain

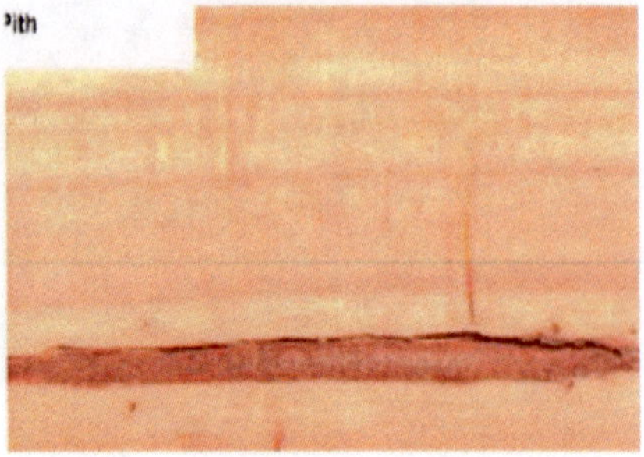

Pith

Plate 4 (continued)

Roller Marks

Snipe

Plate 4 (continued)

Planer Split

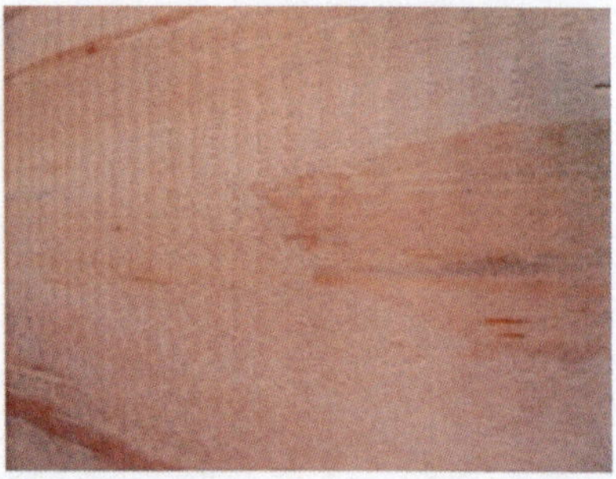

Roller Chatter

Broken Knot

Cutter Marks

Plate 4 (continued)

Plate 5: Furniture Joint Defects

Joints being the weakest-link in the furniture structure must be subjected to much scrutiny to ensure its quality. In this context, the joints must be machined to tight-fit, leading to an optimal glue-line thickness, while facilitating sufficient joint strength build-up. Joint failures can be attributed to either wood failure, or glue-line failure, or both. Very often, failures in furniture is due to poor joints, and the common joint defects are as shown below (Plate 5).

Catastrophic Joint Failure

Joint Failure

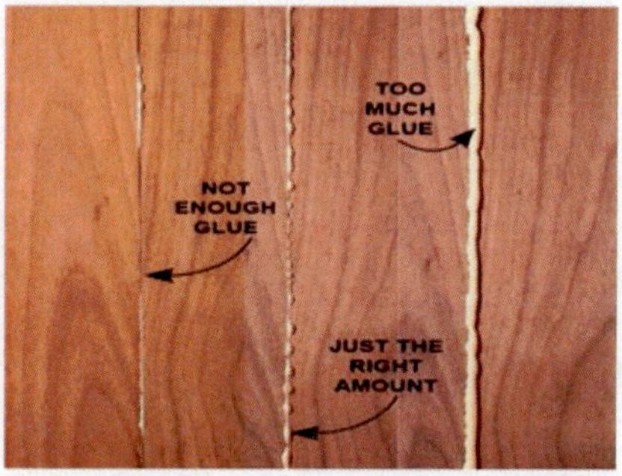

Poor Glue Line

Loose Dowel Joint

Wood Failure at Joint

Plate 5 Furniture joint defects (Courtesy of the Malaysian Furniture Council)

Plate 6: Sanding Defects

Abrasive sanding, unlike conventional machining operation, removes the wood stock through a scrapping action, rather than single chip removal. Hence, if the sanding process is carried out in a non-optimal condition (i.e., too high feed speed, too large stock removal, improper sanding belt use, improper sanding grit sequence, etc.), the following sanding defects may manifest on the sanded wood surface (Plate 6).

Poor Machined Surface

Over Sanding

Roller Chatter

Fuzziness in Sanding

Sanding Belt Chatter

Plate 6 Sanding Defects (Courtesy of the Malaysian Furniture Council)

Plate 7: Coating/Finishing Defects

Wood coating/finishing operation is the cosmetic process in furniture manufacturing. Its implications are far reaching, as it determines the marketability of the piece of furniture. Common coating/finishing defects in furniture can manifest due to wrong choice of coating or finish-type, poor application technique or simply unskilled sprayer. In many instances in furniture manufacturing, the impact of wood coating or finish defects can have very severe economic repercussions, leading to customer rejection of the furniture. Hence, the common coating/finishing defects shown below should be avoided, to ensure the finished furniture is of the highest quality (Plate 7).

Dry Spray

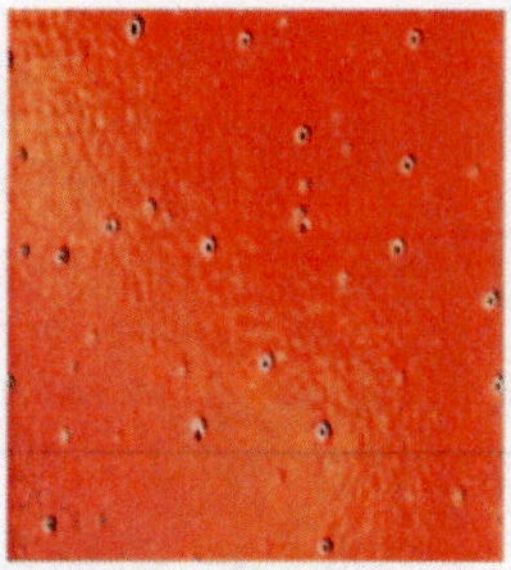

Silicone Poisoning

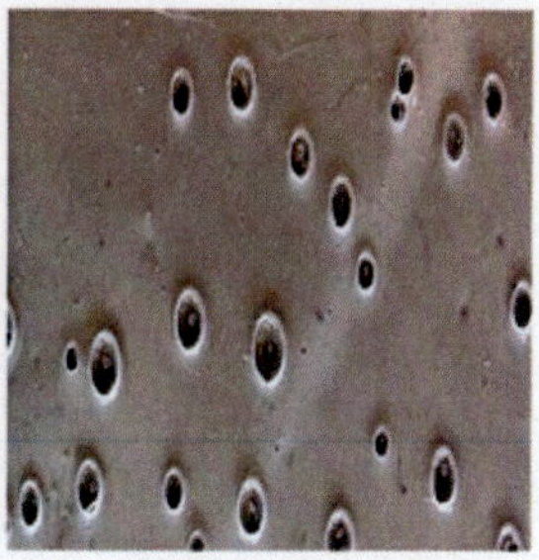

Fish-Eye/Oil Contamination

Flaking

Mold/Fungi

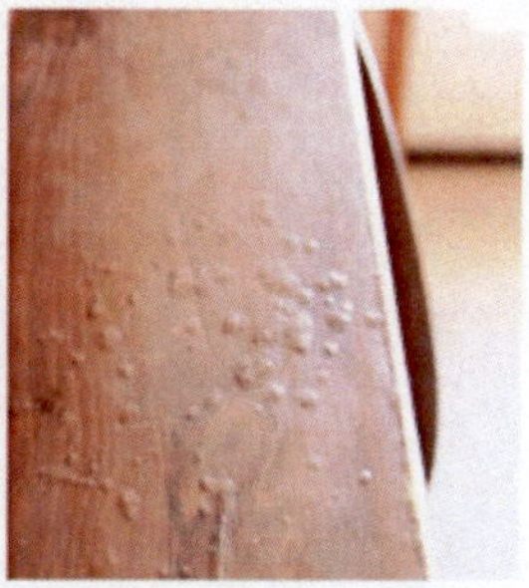

Blister

Wrinkle

Orange Peel

Lifting

Plate 7 Coating/Finishing Defects (Courtesy of the Malaysian Furniture Council)

Blushing

Dirt

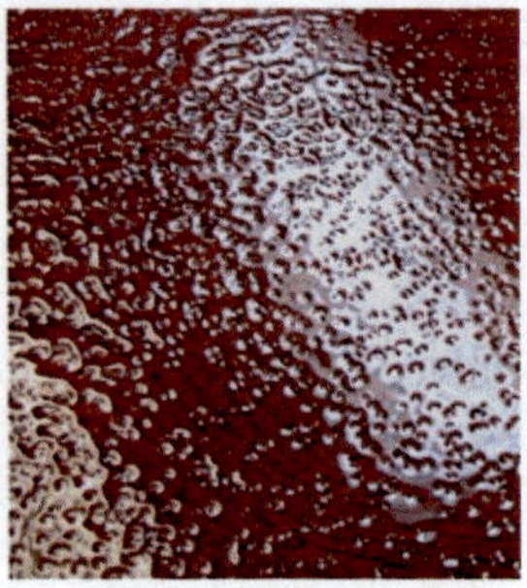

Crawling

Lacquer Sag/Run

Plate 7 (continued)

Index

J. Ratnasingam, *Furniture Manufacturing*, Design Science and Innovation, https://doi.org/10.1007/978-981-16-9412-7